Designing the Reclaimed Landscape

The past decade has been witness to a resurgence of interest in ecologically sustainable thinking about the design and management of reclaimed, post-industrial landscapes. As a result, landscapes that were not previously considered fit for habitation are now being rehabilitated and redeveloped for new uses. *Designing the Reclaimed Landscape* describes new thinking about landscape, and applies new techniques to the task of transforming outdated and disused post-extraction landscapes through design. In the USA alone, there are over 500,000 abandoned mines in need of reclamation and this book provides the first in-depth guidance on this real and pressing issue.

Drawing on the work of the Project for Reclamation Excellence at Harvard's Graduate School of Design and including a wide range of contemporary studies, *Designing the Reclaimed Landscape* outlines the latest design thinking, theory and practice. The book will be a valuable tool for landscape planners, landscape architects and designers, as well as others interested in maximizing the future potential of reclaimed land.

Alan Berger is Associate Professor of Landscape Architecture and Director of the Project for Reclamation Excellence at Harvard University Graduate School of Design. He is the author of *Reclaiming the American West* and *Drosscape: Wasting Land in Urban America*.

Designing the Reclaimed Landscape

Edited by
Alan Berger

Taylor & Francis
Taylor & Francis Group
LONDON AND NEW YORK

First published 2008
by Taylor & Francis
2 Park Square, Milton Park, Abingdon, Oxon OX14 4RN

Simultaneously published in the USA and Canada
by Taylor & Francis
711 Third Ave, New York, NY 10017

Taylor & Francis is an imprint of the Taylor & Francis Group, an informa business

© 2008 Alan Berger for selection and editorial matter; individual chapters, the contributors

Typeset in Frutiger Light by
Book Now Ltd, London

All rights reserved. No part of this book may be reprinted or reproduced or utilised in any form or by any electronic, mechanical, or other means, now known or hereafter invented, including photocopying and recording, or in any information storage or retrieval system, without permission in writing from the publishers.

British Library Cataloguing in Publication Data
A catalogue record for this book is available from the British Library

Library of Congress Cataloging in Publication Data
Designing the reclaimed landscape / edited by Alan Berger.
    p. cm.
1. Abandoned mined lands reclamation. 2. Landscape design. I. Berger, Alan, 1964–

S621.5.M48D47 2008
712—dc22   2007020382

ISBN 10: 0–415–77303–2 (hbk)
ISBN 10: 0–203–93573–X (ebk)

ISBN 13: 978–0–415–77303–4 (hbk)
ISBN 13: 978–0–203–93573–6 (ebk)

To Dorothy Herling
and
Clara Brewster (1917–2006)

Contents

Contributors ix
Acknowledgments xiv
Introduction xvii
Alan Berger

Part I Contextualizing landscape alteration through historic, systemic, and biologic perspectives 1

1 **Valuing alteration** 3
   Frederick Turner

2 **Disturbance ecology and symbiosis in mine-reclamation design** 13
   Peter Del Tredici

3 **Gold and the gift: theory and design in a mine-reclamation project** 26
   Rod Barnett

4 **Mines and design in their natural context** 36
   Dorion Sagan

5 **Ecological succession and its role in landscape reclamation** 42
   Eric D. Schneider

6 **Interrogating a landscape design agenda in the scientifically based mining world** 52
   Belinda Arbogast

Part II Interdisciplinary responses and opportunities in reclamation 61

7 **Science, art, and environmental reclamation: three projects and a few thoughts** 63
   T. Allan Comp

Contents

8 The Wellington Oro mine-site cleanup: integrating the cleanup of an abandoned mine site with the community's vision of land preservation and affordable housing — 77
Victor Ketellapper

9 Building partnerships for post-mining regeneration: Post-Mining Alliance at the Eden Project — 87
Caroline Digby

10 Community-based reclamation of abandoned mine lands in the Animas River watershed, San Juan County, Colorado — 98
William Simon

11 Case studies of successful reclamation and sustainable development at Kennecott mining sites — 105
Jon Cherry

Part III Technology, representation, and information in reclamation design — 113

12 Digital simulation and reclamation: strategies for altered landscapes — 115
Alan Berger and Case Brown

13 Open-pit opportunities: pre-mine design strategies — 125
Alan Berger and Case Brown

14 Reclaiming the woods: trail strategies for the Golden Horseshoe's historic mining roads — 129
Alan Berger and Bart Lounsbury

15 Real-time coal mining and reclamation: the Office of Surface Mining's Technical Innovation and Professional Services program — 132
Billie E. Clark, Jr.

Part IV Future directions and programs in US reclamation policy and law — 139

16 The land revitalization initiative: landscape design and reuse planning in mine reclamation — 141
Edward H. Chu

17 The legal landscape — 154
Robert W. Micsak

Index — 165

A color plate section appears between pages 122–23

Contributors

Editor
Alan Berger is an Associate Professor of Landscape Architecture at Harvard University's Graduate School of Design where he teaches courses to landscape architects, urban designers, planners, and architects. He founded and directs P-REX, the Harvard University Graduate School of Design's Project for Reclamation Excellence, a multi-disciplinary research effort focusing on the design and reuse of post-mined landscapes. His other books include the award-winning *Reclaiming the American West* (New York: Princeton Architectural Press 2002), and *Drosscape: Wasting Land in Urban America* (New York: Princeton Architectural Press 2006). Berger is a Fellow of The American Academy in Rome.

Contributors
Belinda Arbogast has over 25 years of professional experience as a chemist and physical scientist with the U.S. Geological Survey. Belinda's current research merges earth science information in landscape assessment with societal impacts for land use planning of industrial minerals development. Her map, *Evolution of the Landscape along the Clear Creek Corridor, Colorado – Urbanization, Aggregate Mining, and Reclamation*, was awarded a first place in the 2003 National Association of Government Communicators' Blue Pencil Competition. Ms Arbogast earned her Masters of Landscape Architecture from the University of Colorado, where she received its design excellence and outstanding graphics awards, as well as the national award of honor from the ASLA. She has lectured internationally, published technical papers, and has experience in both the design and geotechnical fields.

Rod Barnett is Associate Professor of Landscape Architecture at Unitec Institute of Technology, Auckland, New Zealand. His PhD investigated the discourse of nature in landscape architecture with a particular emphasis on the roles of emergence and disturbance in landscape architectural design. He is founder and director of the Unitec Landscape Unit which provides research and consultancy services to public and private organizations. The Unit has a special interest in nonlinearity and adaptivity – characteristics of self-organizing systems – in the design and management of landscape systems. Many landscape ecologies are dissipative structures. On the one hand they rely on disturbance, and on the other on waste in order to maintain their dynamism, flexibility, and organizational integrity.

Rod's current research interests include the application of disturbance ecology operations to the re-design and re-building of Pacific Island settlements after cyclonic disturbance, and the use of nonlinear landscape design propositions (based on open-endedness, unpredictability, and self-organization) to actualize landscape interventions in the marginalized wastelands of the Auckland region.

Contributors

Case Brown works as a Research Associate at the Harvard University Center for the Environment. He works with several faculty members across the university, researching a broad array of topics from reclamation of altered sites to trends in urbanization, economics, and natural phenomena. While researching these areas, he utilizes the digital environment, from GIS to 3D visualization software, to better represent spatial and numerical data. He was named the 2006 recipient of the Charles Eliot Traveling Fellowship from the Landscape Architecture Department at the Harvard Graduate School of Design. With this fellowship, he intends to document and analyze the process of militarization on the border landscapes of the United States. He holds a BA from Duke University in Art History.

Jon Cherry, P.E. is manager of Environment and Governmental Affairs, Kennecott Minerals Company. Mr Cherry is a registered professional engineer in the field of environmental engineering with 17 years of experience in the mining industry specializing in environmental management and design. At Kennecott Minerals Company he is the project manager for the Eagle Project, a green field nickel discovery located in the Upper Peninsula of Michigan that is in the process of being designed and permitted as the only primary nickel mine in the United States. Prior to tackling the Eagle Project, Mr Cherry spent 12 years at Kennecott Utah Copper managing compliance and permitting issues as well as various remediation projects.

Edward H. Chu is Acting Director, Land Revitalization Office, U.S. Environmental Protection Agency's (EPA) Office of Solid Waste and Emergency Response.

The Land Revitalization Office was created in September 2004 with the long-term goal to return all contaminated and potentially contaminated properties to beneficial and environmentally responsible uses for America's communities. Ed is currently leading the development of a long-term strategic work plan on land revitalization. Prior to joining the Land Revitalization Office, Ed led the Regulatory, Economics, and Analysis Group at the U.S. EPA's Office of Children's Health Protection. Ed's primary responsibility was to help people (both inside and outside the government) find and use regulatory, economic, and statistical information to better understand and protect children's health from environmental contaminants. Prior to his work on children's health issues, Ed worked in EPA's Indoor Environments Division where he worked on privatizing the Radon Proficiency Program. Prior to joining EPA, he was a consultant specializing in environmental and energy issues. Ed received degrees from the University of Michigan and Michigan State University. He grew up in Honolulu, Hawaii and now lives in the Washington D.C. area. Living with their two daughters (2 and 5 years old), Ed and Karen spend much of their time on family activities.

Billie E. Clark, Jr. has held numerous positions with the Department of the Interior over the past 26 years. He has supervised and managed the Interior's Federal and Indian lands energy and minerals programs while employed by the Office of Surface Mining (OSM), Minerals Management Service (MMS), and the United States Geological Survey (USGS). He earned a BS Civil Engineering, New Mexico State University in 1976. Mr Clark served as OSM manager for Indian lands and Federal programs permitting for 12 years in Denver, Colorado. In this capacity he managed and oversaw permitting and National Environmental Policy Act (NEPA) functions.

Since 1999 Mr Clark has managed the OSM's national Technical Innovation and Professional Services Program (TIPS) and Western Region IT Operations. In cooperation with State and Tribal regulatory and reclamation agencies, as well as Office of Surface Mining offices nationwide, TIPS provides scientific and engineering software and hardware tools for Federal, State, and Tribal experts to do their jobs faster and more efficiently. TIPS is an important tool for regulators in carrying out the Surface Mining Control and Reclamation Act of 1977. Under Mr Clark's leadership TIPS now provides 26 high-end scientific and engineering applications; mobile computing hardware and software solutions; geospatial solutions (including Geographic

Information Systems and remote sensing); 3D modeling; 25 specialized software training courses; and technical assistance.

T. Allan Comp holds a PhD in history of technology and economic history (Hagley Fellow: University of Delaware) with a long commitment to cultural resources, community engagement, and environmental recovery. Allan Comp is currently employed full-time by the Office of Surface Mining where he is focused on supporting the efforts of volunteer watershed groups working for the recovery of the Appalachian Coal Country from a century of pre-regulatory exploitation and neglect.

Allan Comp and his AMD&ART project (which he directs as a volunteer) won the first national 2005 EPA Brownfields Phoenix Award for community impact on mine-scarred lands, the Pennsylvania Environmental Council Green Design Award, and an article in *Landscape Architecture* in October, 2005 (pp. 96–115). His work in building partnerships in coal country was recognized by the Secretary of the Department of the Interior in 2005 with her *National 4-Cs Award* and by a 2004 *Environmental Achievement Award* from Interior for his OSM/VISTA Team. The Wildlife Habitat Council, the Alliance of Artist's Communities, and the National Council of RC&Ds all singled out Allan Comp's work as an innovator in approaches to reclamation and partnership building.

Allan Comp served as an intermittent Artist in Residence at the Sitka Center for Arts and Ecology in 2004–5, leading his *Crowley Creek Collaboration* project, bringing artists and scientists together to develop a new adaptive approach to river restoration and community engagement. He received a *Bridge Residency* at the Headlands Center for the Arts in 2000 for his work as an "artist/thinker bridging to other disciplines" and was part of a small team awarded the national EDRA/*Places* magazine Place Planning award the same year. He authored the first Brownfields Pilot Demonstration Project awarded to a coalfield watershed, followed that success with two more in Tennessee and West Virginia, started both the Office of Surface Mining Summer Watershed Internship Program and the OSM VISTA Initiative, and wrote *Hope and Hard Work*, a booklet funded by EPA and published by the Canaan Valley Institute, celebrating the work of nearly two dozen coal-country watersheds.

Caroline Digby was appointed as Development Director by the Eden Project in July 2004 to turn the Post-Mining Alliance from concept to reality. She joined Eden from the International Council on Mining and Metals (ICMM) where she was Programme Director responsible for community and social development and health and safety activities. Prior to ICMM, Caroline was the Research Director for the Mining, Minerals and Sustainable Development project at the International Institute for Environment and Development. She has a BA in Economics from Trinity College Dublin, MA in Economics from the University of British Columbia, and an MSc in Environmental Assessment and Evaluation from the London School of Economics.

Victor Ketellapper works for the U.S. Environmental Protection Agency (EPA) in Denver, Colorado managing the investigation and cleanup of Superfund sites. The majority of the sites he has managed have been mining related. Sites managed include the Summitville Mine and French Gulch. While with the EPA, he has been actively involved with the development of policy, guidance, and practice of incorporating future land use planning into EPA cleanup actions. Prior to joining EPA, Victor worked as a consulting engineer. He has Bachelor's Degrees in Economics and Geology and a Master's Degree in Environmental Engineering. He is a professional engineer, registered in the State of Colorado.

Bart Lounsbury is a Beagle Foundation–Harvard Law School Fellow at the Natural Resources Defense Council in Santa Monica, California. He grew up in the Maine woods, cross-country skis in one hand and a mountain bike in the other. Trails have been his passion ever since. After college, Bart indulged his outdoor zeal, attacking invasive weeds and assisting visitors as a park

Contributors

ranger at Point Reyes National Seashore and Canyonlands National Park. During graduate school, Bart specialized in environmental law and planning and wrote his thesis on the legal ramifications of mine remediation. He received a Switzer Foundation Environmental Fellowship for his graduate work and an American Society of Landscape Architects' Award of Honor for his Breckenridge trails project. Bart holds degrees in French literature (AB), urban planning (MUP), and law (JD) from Harvard University.

Robert W. Micsak is currently the Executive Vice President and General Counsel for CBS Outdoor. Prior to this he served in a number of senior executive roles with Gulf Oil, Chevron, Burlington Resources, Minorco, and AngloGold. In these positions he was responsible for hundreds of mining projects, from initial land acquisition to final project approval, including final reclamation planning and implementation. His responsibilities included assuring that the project plans and permit conditions were implemented in the final construction and ongoing operations at each site. He also managed the litigation that arose in the course of the project approval process. He has worked around the globe, across North and South America, Australia, Africa, and Europe on a variety of mineral development projects, including metals, energy minerals, including coal and oil shale, aggregates, and a variety of industrial minerals. Projects in which he has been involved have occurred on a broad range of private, state, tribal, and federal lands and his project teams obtained some of the fastest project approvals in the industry.

His work has involved virtually every major environmental and reclamation law and regulation in the United States, including a host of land use laws associated with mineral development, including the National Environmental Policy Act, Clean Water Act, Clean Air Act, Endangered Species Act, Resource Conservation and Recovery Act, Comprehensive Environmental Response Compensation Liability Act (Superfund), Toxic Substances Control Act, Surface Mining Control and Reclamation Act, Federal Land Policy Management Act, National Historic Preservation Act, Forest Service Organic Act, Mining Law of 1872, Common Varieties Act, Federal Coal Leasing Act, and the Mineral Leasing Act of 1920. He has testified in the US Congress, in various state legislatures, and taken leadership roles in industry association groups, lobbying on a variety of issues related to environmental and mineral development issues.

Dorion Sagan is author of some fourteen books. His writing has appeared in *The New York Times*, *Wired*, *After Hours*, *Pabular*, *Cabinet*, *Clean Sheets*, *Tricycle*, *The Ecologist*, *Natural History*, *The New York Times Book Review*, *Omni*, and *The Times Higher Education Supplement*. His latest project, with Montana thermodynamicist Eric D. Schneider, is *Into the Cool: Energy Flow, Thermodynamics and Life* (Chicago, IL: University of Chicago Press 2005).

Eric D. Schneider works at the intersection of biology and physics and uses thermodynamics to examine these connections. He has his PhD – from Columbia University – in marine geology. He is Director of a global ocean floor analysis laboratory for the United States Navy as well as Director of the United States Environmental Protection Agency's National Marine Water Quality Laboratory. He is a Senior Scientist: National Oceanic and Atmospheric Administration and the recent coauthor of the book *Into the Cool* (Chicago, IL: University of Chicago Press 2005).

His research centers on the study of energy flow through natural systems. Under the right constraints, energy flow can lead to organization and complexity. He is interested in the emergence of processes in inanimate systems such as hurricanes, as well as the study of highly complex systems such as life. Schneider has used ecosystems as laboratories to help understand the emergence and development of complex systems. Ecosystems are dynamic systems that organize and develop in predictable patterns. The very process of ecosystem succession is a manifestation of the second law of thermodynamics. Successions are restorative processes.

William Simon is the owner of a consulting and ecological restoration company located in Hermosa, Colorado. He has worked to restore numerous abandoned mine lands, providing design, construction, and monitoring. He has had significant successes during 27 years of high mountain mine land restoration efforts and has two decades of experience in stream restoration, fish habitat development, and wetlands treatment development.

Thirteen years ago, realizing a need to encourage voluntary cleanup of abandoned mine lands, Mr Simon helped initiate the Animas River Stakeholders Group. He has held the position of Animas Watershed Coordinator since 1994. Mr Simon has given numerous presentations on water quality issues, mine land restoration, and collaborative approaches to resource management at conferences and educational events throughout the western United States. As editor and co-author, he recently completed the Animas Watershed Use Attainability Analysis considered one of the most complete and comprehensive documents that characterizes the chemical, biological, and physical condition of a large, mine impacted watershed. The report prioritizes sites for remediation, provides recommendations for implementation of specific remediation technologies at each site, and was used to set new goal-based water quality standards and 29 TMDLs throughout the watershed. Years of remediation have recently been followed by a basin-wide analysis which demonstrates significant chemical and biological improvements in some, but not all, streams.

Mr Simon has a BA in biology from the University of Colorado and completed the PhD program in evolutionary ecology at the University of California, Berkeley. He is also a former San Juan County Commissioner and a 36-year resident of the Animas Watershed. He maintains a small farm and orchard and is an avid skier, rock climber, hiker, and white water boater.

Peter Del Tredici has worked at the Arnold Arboretum of Harvard University since 1979, in a variety of capacities as a plant propagator, Editor of *Arnoldia*, Director of Living Collections, and, most recently, Senior Research Scientist. Since 1984, he has been Curator of the famous Larz Anderson collections of bonsai plants, housed at the Arboretum. Dr Del Tredici has been a Lecturer in the Department of Landscape Architecture since 1992, with an interest in urban ecology. He is the winner of the Arthur Hoyt Scott Medal and Award for 1999, presented annually by the Scott Arboretum of Swarthmore College "in recognition of outstanding national contributions to the science and art of gardening."

Dr Del Tredici has worked on various aspects of both botany and horticulture over the last 25 years. His interests are wide ranging and include such subjects as plant introduction from China, the root systems of woody plants, stress tolerance in urban trees, and the cultural and natural history of the Ginkgo tree.

Frederick Turner is Founders Professor of Arts and Humanities at the University of Texas at Dallas, is a poet, a cultural critic, a philosopher of science, an interdisciplinary scholar, an aesthetician, an essayist, and a translator. Born in England in 1943 to the anthropologists Victor W. and Edith L. B. Turner, he grew up in Central Africa and was educated at Oxford University. His dissertation, "Shakespeare and the Nature of Time," was published by Clarendon Press. He has taught at UC Santa Barbara and Kenyon College, where he co-edited the Kenyon Review. A winner of the Levinson Poetry Prize and the Milan Fust Prize, Hungary's highest literary honor, he is the author of twenty-two books, including *Natural Classicism: Essays on Literature and Science*; *Genesis: An Epic Poem*; *Rebirth of Value: Meditations on Beauty, Ecology, Religion and Education*; *Beauty: The Value of Values*; *April Wind and Other Poems*; *Foamy Sky: The Major Poems of Miklos Radnoti* (with Zsuzsanna Ozsvath); *Biopoetics: Evolutionary Explorations in the Arts* (essays by various hands, edited with Brett Cooke); *The Culture of Hope*; *Hadean Eclogues*; and *Shakespeare's Twenty-First Century Economics*.

Acknowledgments

Neither the Project for Reclamation Excellence (P-REX) nor this publication would exist without the support of colleagues at many institutions worldwide. I especially want to thank the seventeen authors in this collection, whose unique perspectives open the imagination onto the possibilities of reclamation thinking and design. I am indebted to my colleagues in the Department of Landscape Architecture at the Harvard Graduate School of Design for providing the platform from which to integrate and disseminate reclamation knowledge into the design community. I am also grateful to Niall Kirkwood, who supported P-REX's larger design-research agenda during his tenure as chair of the Department of Landscape Architecture, as did his predecessor, George Hargreaves. Conversations with Richard Forman, Jerold Kayden, Martha Schwartz, John Beardsley, Richard Zeckhauser, James Corner, Charles Waldheim, Gene Bressler, Jane Amidon, Chip Sullivan, Michael Jensen, Gale Fulton, Tim Beard, Kelly Shannon, Evan Douglas, Tom Buresh, Robert Adams, Jason Young, Juan Rois, Dan Hoffman, Lars Lerup, Albert Pope, and Ken Smith were invaluable in broadening the reclamation agenda. A special thanks to Alan Altshuler and Peter Rowe, two former Deans of the GSD, for supporting P-REX and its program during their tenures. In addition to my school, Dan Schrag, director of the Harvard University Center for the Environment, deserves my grateful acknowledgment for co-sponsoring and tirelessly supporting my work, as well as for introducing me to faculty across the university who are concentrating on environmental concerns. I also thank Roger Kennedy for his contribution at the P-REX2 symposium, which is not included in this collection. P-REX events, which are all free and open to the public, would not be possible without the generous co-sponsorship of the following organizations and people: the U.S. Environmental Protection Agency; Carol Raulston, National Mining Association; Tiffany & Co. Foundation; Town of Breckenridge, Colorado; Marcelle Shoop, Rio Tinto Group and Kennecott; and Roger Argus, Tetra Tech. Color funding was provided by Harvard University Center for the Environment, Rio Tinto Group and Kennecott.

The work of P-REX is catalyzed by my students and researchers who have participated in and helped shape its programs and projects over the years. These P-REX research associates include Case Brown, Richard Kennedy, Pippa Brashear, Bart Lounsbury, Gena Wirth, and Scott Melbourne. I additionally thank my students Elizabeth Fain, Sarah Cowles Gerhan, and Megumi Aihara, who have worked in various capacities for P-REX. I am grateful to Grace Kulegian and Aimee Taberner for helping organize symposia logistics. Thanks to Lauren Neefe for her editorial prowess early in the manuscript's development. Finally, I thank my two editors at Routledge, London, Ellie Rivers and Kate McDevitt, for enthusiastically supporting this book and guiding it through an incredibly fast production schedule. This book would never have been accomplished without the constant support of my family, Elaine and Leah.

P-REX

Harvard University Graduate School of Design
PROJECT FOR RECLAMATION EXCELLENCE

Introduction

Project for Reclamation Excellence: P-REX

Mining is an integral part of the modern American landscape. The western United States alone has more than five hundred thousand abandoned and active mines, covering millions of acres and tens of thousands of square miles. At current production rates, most of the mineral and energy resources will be mined out of federal land in the American West during the twenty-third century. Although the total acreage is difficult to predict, a vast new post-mined landscape, approaching the scale of a hundred thousand square miles, will be created in the wake of US mining alone. This landscape will need to be redesigned, reclaimed, and reprogrammed for post-mine land uses (Berger 2002).

After years of conversations with colleagues in both the design and the mining worlds, it became clear to me that reclamation, design, and the environment could mutually benefit one another. A forum for cross-fertilization of practices, scholarship, and applied research needed to be created with the singular goal of mandating landscape change on the vast scales at which extractive industries were already permanently altering landscape systems. I founded the Project for Reclamation Excellence, or P-REX, to forge these new cross-fertilizing relationships among the design professions, the extractive industries, and the regulators that oversee mining and reclamation activities. P-REX is now a project of the Department of Landscape Architecture at the Harvard University Graduate School of Design. Under my direction, the project partners with like-minded groups to implement a multidisciplinary research program aimed at integrating design with the reclamation of natural-resource extraction sites.

The greatest motivation for P-REX lies in the recognition that domestic and international natural-resource extraction and development will rapidly increase over the next quarter millennium, as industrial production and consumption globalize. Compounding this phenomenon is the fact that few places outside North America, the European Union, and Australia have fully developed standards or programs for reclaiming mined lands. To understand the extent and diversity of this situation, as well as the opportunities for design within it, consider these facts:

- When the P-REX1 symposium was held at Harvard, in fall 2004, gold was selling at a 16-year high of $430 per ounce. By the time P-REX2 was held, in spring 2006, both gold and silver futures recorded a 25-year high, with gold selling at more than $610 per ounce and copper hitting its all-time-highest price.
- Almost every urbanized area with a population of more than fifty thousand people has at least one gravel or aggregate mine for producing road base and concrete.

- By 2012, China, India, and the United States combined are planning to build more than 800 new coal-fired power plants, which will burn about 900 million extra tons of coal each year. China is on track to add 562 coal-fired plants, India could add 213 such plants, and the USA 72. The result will be several hundred thousand square miles of new surface-coal-mine disturbance, all of which will eventually need concurrent landscape reclamation.
- Nonfuel mineral production is increasing as a result of human consumption. Thirty percent of all aluminum production, for example, is for automobiles and transportation uses. The prices of copper and gold were selling at a 16-year high at the time of this publication. The USA remains the second-largest gold-producing nation; just 39 mines yield 99 percent of the country's gold. Mergers and acquisitions are the order of the day, leading companies to seek larger targets, thereby creating larger landscape disturbances.
- Ninety-seven percent of the US specialty- and precious-metal-producing areas are in the Intermountain West, a region that has sustained rapid population growth for the past 20 years and shows no signs of slowing.
- On January 11, 2002, the President of the United States signed into law the Small Business Liability Relief and Brownfields Revitalization Act. Since its inception, this law has expanded the definition of Brownfields to include mine-scarred lands, making these properties eligible for the benefits of the Environmental Protection Agency's (EPA's) Brownfields program.
- The mining world and its investors value reclamation now more than they have in the past. This change of attitude was highlighted in 2003 when ten large banks adopted voluntary guidelines of sustainability for the mining projects that they finance; the guidelines are based on the policies of the World Bank and the International Finance Corporation. Among the banks that adopted the guidelines were Barclays, Citigroup, and Credit Suisse, which together underwrote about 30 percent of worldwide project loans in 2002. The International Council on Mining and Metals has also been expanding its social and environmental programs. The mining world now realizes that, beyond the lawful requirements of reclamation, exceptional design and "good neighbor" policies make it easier to process permits for new mining elsewhere.

In the United States, federal and state laws now require the return of all mines to a beneficial post-mine land use. Across the entire American landscape, but especially in the American West and South, where rapid population growth, horizontal urbanization, and in-migration over the past 15 years have expanded urbanized areas, more people live in closer proximity to mined sites than ever before. As natural-resource extraction, urbanization, economic development, and living interface in more complex ways, designers become agents in deciphering how to act on the resultant landscapes. P-REX can not only become the epicenter for global leadership on reclamation issues regarding the design and reuse of post-mined landscapes, but it can also serve as an incubator for design research on any altered landscape. Given the magnitude and scale of permanent landscape destruction, designers can take advantage of mined and reclaimed sites to see and plan landscape systems in new ways.

Why are mined landscapes important to the design fields? Mined sites enable designers to speculate over a landscape that is not bound by, nor indebted to, historical filters, aesthetic traditions, or strict contextuality. Within this rubric, reclamation is not solely a "landscape" or "engineering" issue. It is a large-scale design issue that affects environmental systems and the life supported by them. Most importantly, reclamation can act as a laboratory for experimentation.

The chapters in this collection were gathered from several sources. The greater portion of them were first presented at two symposia held at the Harvard Graduate School of Design: "P-REX1: Projecting Reclamation in Design" (2004) and "P-REX2: Reclaiming the Future" (2006) (see Figures 0.1A, 0.1B, 0.1C, and 0.2). These symposia convened a total of 22 speakers, addressing a rich array of reclamation topics and concerns. The participants included federal officials, poets, miners, scientists, philosophers, landscape architects and other designers,

Introduction

Figure 0.1A "P-REX1: Projecting Reclamation in Design" symposium and exhibition, held at Harvard University Graduate School of Design, December 1, 2004–January 17, 2005.

Figure 0.1B "P-REX1: Projecting Reclamation in Design" symposium and exhibition, held at Harvard University Graduate School of Design, December 1, 2004–January 17, 2005.

Alan Berger

environmental lawyers, historians, and community activists, in addition to the hundreds of audience members who interacted with these speakers through panel discussions (see Figure 0.2). Further sources of the ideas here collected are my teaching, research, and fieldwork on reclamation sites, as well as ongoing research with the Superfund Program and US/German Bilateral Brownfields and Brownscape Group of the U.S. Environmental Protection Agency. This work has led to several design projects, two of which are described in the following pages. Another

Figure 0.1C "P-REX1: Projecting Reclamation in Design" symposium and exhibition, held at Harvard University Graduate School of Design, December 1, 2004–January 17, 2005.

Figure 0.2 "P-REX2: Reclaiming the Future" symposium, held at Harvard University Graduate School of Design, April 21, 2006.

source of inspiration embedded in this collection is the constant encouragement and critical feedback I have received from numerous colleagues.

This book is not intended to specify a univalent disciplinary role for landscape architects – or other designers – in mining and reclamation. On the contrary, these chapters suggest that reclamation, as a unique form of landscape production, offers designers a substantial opportunity to expand their intellectual concerns and scholarship in the areas of landscape disturbance, renewal, design, and of reoccupation of synthetic space and ecology. By learning from mining and reclamation, landscape architecture could redirect and expand its influence in the areas of sustainability, site evolutionary strategies, and new ecological thinking.

The word *reclamation* captures at once the temporal and ethereal issues of designing with altered landscapes: one must think backward in time in order to project forward in time; one must change old trajectories in order to assemble new solutions. Moreover, one must intercept time – as if it were, in a design sense, suspended – just long enough to rethink the roles of human agency on evolution and the overlay of human intentions on the landscape that alters nature's methods of time management and dynamic change. If the future could really be reclaimed, one might shift design thinking and practice toward crisis management, so that landscape catastrophes of various sorts could be avoided: for example, seismic design of infrastructure, flood-control measures, so-called invasive species and pest infestations, or the productive reuse of environments that will be permanently altered by mining enterprises. If the idea of designing with altered sites is considered broadly, then the notion of replacing time, or designing with merely the palette of historical accuracy, such as in restoration, is anathema. That is, as designers, we are choosing to reclaim, rather than restore, when we project into the future, which is why the topic of reclamation is so vital to today's and tomorrow's *design discourse*.

Another argument that runs throughout the following essays involves methods of integrating, through design, the reclamation of natural-resource-extraction sites into the ongoing needs of local and regional communities. The reality of reusing mined sites requires cross-disciplinary thinking and knowledge, involving no less than public policy, law, ecology, biology, economics, political activism, engineering, and a variety of landscape-architectural outlooks and design fields.

The ideas in this collection pose many more questions than answers. As with any new cross-disciplinary approach, the primary objective is the formation of knowledge that can redirect our current ways of thinking about reclamation – to allow for the mining world to begin a new discourse with the design world, and vice versa. In a globalized economy of higher demands for natural-resource extraction, we can no longer afford to think of designing with altered terrain as a "dirty" alternative to high or boutique design, nor as a merely artistic, sculptural, or intellectual indulgence or self-referential process. The readers of this book need to rise above this situation and invest in the idea that, as students, scholars, practitioners, and stewards of global landscape concerns, you are the front lines of knowledge creation, and incremental disciplinary reformulation follows in your wake. Reclamation will eventually be absorbed into landscape-architecture practice, but not in the form of planting plans, parking-lot designs, screening techniques, and berm making. Landscape architects must engage reclamation at its origins – by embracing and valuing alteration – as much as we feel compelled to resist and deny that change is happening.

Bibliography

Berger, A. (2002) *Reclaiming the American West*, New York: Princeton Architectural Press.

Part I
Contextualizing landscape alteration through historic, systemic, and biologic perspectives

Chapter 1
Valuing alteration

Frederick Turner

An interesting point came up in a recent conversation with two students on the subject of this volume. I said that I wanted to identify what makes good reclamation. My students, who are brothers, had grown up in rural Kentucky, where strip mining had been extensive. They told me about an area near their home which had been abandoned after the stripping. This place was for them a wonderland, with little lakes and ponds that were full of fish, lots of miniature hills, and a switchbacked isthmus between two ponds that was perfect for riding bikes across. The place was full of wildlife – birds and mammals and butterflies. A whole community came there to fish and play and ride.

Then, in due course, the reclaimers came and graded it all, leaving one boring lake and reducing this magic place to a prosaic field with a dirt road leading to a farm. Something had gone amiss with the task of reclamation. How do we distinguish between good and bad reclamation?

The work of reclamation began as a practical need, a public-health necessity, a legal problem, and a technological challenge. This volume inaugurates and celebrates a second phase in the work; the point in any craft or occupation or activity at which the need arises to look at itself, to assess its performance, to refine its methods, to develop measures and techniques by which its results can be examined. The editor of this volume is making an admirable start, and it will indeed have been a success if it does no more than provide the foundations for answers to such questions as what is reclamation?, how is it done?, and what are its constraints (legal, technological, geological, biological, etc.)?

In this context, it may still be premature to embark on a more philosophical exploration of the subject. Yet sometimes it helps, when climbing a mountain, to have an aerial photo of the whole, even if that photo is obscured by cloud and distorted by perspective. Alan Berger has shown us how valuable an aerial picture can be.[1] Certainly, a good map of our subject must wait until we climbers have done a survey on the ground. If my analogy holds, however, even a poor picture of the peak – the goal – and of its approaches may be valuable; and, with regard to reclamation, that poor picture would be provided by preliminary answers to some big value questions: What *should* reclamation be? What is good reclamation?

In order to answer these questions, we must first answer others. "Good for what? For whom?" It seems to me that three main "stakeholders" have a claim. The first two are obvious, or so it would appear: good for human beings and good for the rest of nature. But even to put the issue in this way is already to raise problems. Are human beings part of nature or not? If not, then we must adopt some theory by which human beings were somehow injected into nature from some sphere that is outside of nature, from some unnatural sphere. The theory of evolution is one of the foundations of the biological science and technology that reclaimers must use daily, if only implicitly, in their work: in considering, for instance, geological succession, immunity to pesticides, species extinction, and competition between native and exotic species. Only if we abandon the theory of evolution – and with it the basic assumptions that make reclamation possible – can we assume a radical distinction between humans and the rest of nature. We are another animal species that evolved, say the evolutionary biologists. So we must assume that humans are part of nature. If humans are as natural as anything else, how do we examine the claims of one part of nature (humans) with respect to those of others (e.g. rare native fish or insects)? Is a forest of humans – a suburb – more or less valuable than a forest of conifers? If there are lots of suburbs but only a few of that species of conifer, does the equation change?

Even if we give a priority to the good of nature as a whole, what criteria of goodness do we use? Should we be trying to maximize the biomass of a place? Or its *living* biomass? Or its richness in species? Or its contribution to the world's species richness? Or its genetic variety? Or its population of higher organisms? Or of organisms with higher nervous systems? Or the richness of the cultures, animal or human, that inhabit it? Or – to change the criteria in other ways – the self-sustainability of the place? Its inheritance of former ecological or genetic patterns? Its efficiency in using solar energy? Its independence from the resources of other parts of the planet?

Each criterion has its own justifications. Sheer biomass has the advantage of not putting us humans in the position of judging which biological material outranks which, thereby giving us the top position in a hierarchy of judgment. *Living* biomass avoids the definitional problems involved in the criterion of biomass: is a caddisworm's jacket of gravel part of its body, or not? Is a termite's nest biomass? What about dead heartwood? My artificial hip joint? The undigested material in an owl's stomach? Richness in species is, for obvious reasons, our common ecological measure. Contribution to species richness remedies the logical incompleteness of the criterion of species richness. Genetic variety speaks more directly to the health and adaptiveness of a living biome. Higher organisms act not only as a "canary," signaling ecological soundness lower down, but also as indexes of several other factors cited as evidence of environmental value; and this criterion avoids the problem of overvaluing, for instance, a landscape that consists only of a billion species of moss and lichen, with wildly various genetic codes. The formula "organisms with higher nervous systems" sharpens the vague term "higher organisms" and also implies a criterion of what we might call "epistemic value": a landscape observed by sentient inhabitants may be worth more than a landscape with no observers. In the same spirit, the cultural criterion of systemic value is sharper still: social animals that can share their observations may be better observers than solitary ones.

Self-sustainability has the advantage of appealing to the sense of peace and balance that we associate with nature. The inheritance of past patterns of life has scientific value. Efficiency in converting solar energy is more easily measurable by physicists, and gets us out of the philosophic tangle already manifest in this discussion. And "independence" has a nice ring of robustness about it.

Each priority would yield different policies. The biomass of a climax forest, in which most of the carbon is sequestered in dead heartwood, might be outranked in the "living biomass" criterion by other biomes, such as a swamp or a fish farm. But should we burn down climax forests to make room for more biodiverse prairies and savannas? A botanical garden might have more species yet – and more endangered ones. A suburb might have more genetic diversity, more nervous systems, more culture. A few species with rich and complicated genomes of many strains might have more genetic variety than a large number of monoclonal species. A eutrophicated pond might use solar energy very efficiently indeed. Barren rock might be very self-sustaining. Or should we be trying to create Jurassic Parks? If we accepted all the priorities I have mentioned, how should we rank them or weight them with respect to one another? Upon what principles? Principles derive from a deep understanding of something.

The question then becomes: what views of nature and of human beings will best serve good reclamation?

There are ways of avoiding some of the knotty philosophical problems involved in this approach. One rule of thumb has been the policy that reclamation should leave a place as close as possible to how it was found. Of course, we always fall short. If this is our goal, then we will always fail. The only difference our work might make would depend on how badly we fail. On purely psychological grounds, this goal would seem to provide a poor standard for rallying enthusiastic recruits to a much-needed profession! The often-used term "remediation" implies the restoration of health to something that was sick. But if health is defined as the status quo ante, the situation that prevailed before the alteration, then the plateau of Arizona was a healthier place before the Colorado incised the Grand Canyon in it, the devastated slopes of Mount St. Helens were healthier than the forests and meadows that have since grown up there, and the frozen rock beneath the glaciers of the last Ice Age was healthier than the mixed deciduous forest of the upper Midwest. So another question, which must accompany our question about human beings and nature, is: can reclamation go beyond remediation?

The term "remediation" is a good one, though, in its sense of healing, of organic process. Broken bones can heal to become stronger than before. To embark on physical therapy in order to bring back wasted muscles – as I have had to do a few times after surgery – is to begin a process of micro-tearing the muscle fibers, forcing the body to heal the tiny lesions with sturdier patches, the cells then warned of the greater challenges they must expect. When athletes train, they do the same thing, only to an already perfectly healthy body. Can we heal our landscapes in such a way as to improve on their original states? Indeed – to continue the sickness-and-healing metaphor – do we not often find that a period of grave or even life-threatening illness and convalescence opens up to us entirely new perspectives in our lives, enables us to pass through one of the maturational metamorphoses that enliven our souls? Can alteration of a landscape have the same effect, providing the landscape with a destiny and a role that are grander than its original ones?

But is nature itself capable of such strange growth? Much of our basic thinking about reclamation – getting back something that was gone, being unsatisfied with anything but what we had before – assumes that nature (presumably untouched by humans) was perfect before, and that its cycles and harmonies are, or ought to be, eternal. It is easy enough to point out that, if this were so, nature made a big mistake in evolving human beings; and if nature can err in one place, it can err in others, and is not as perfect as we thought. We would also have to explain those episodes in the Earth's history when perfectly natural megameteor collisions or eruptions wiped out two-thirds of the planet's species and plate tectonics and ocean currents created a snowball planet, with glaciers down to the equator. The word "nature" itself derives from the Latin word for being born, "nascere," "natus," and the Greek word "physis," which gave us "physics" and "physical," means "growth," or "production." Both the Latin word and the Greek word imply a dynamical process that can create new forms and beings, a paradoxical concept of naming something that *is* in terms of "is"'s "isn'ting" – the process of change and passing away and replacement – for whatever is born must die, and whatever grows thereby subjects itself to decay.

If we see nature as essentially dynamic, open-ended, radically evolutionary, irreversible – and humans as part of its process – we may begin to have a basis for evaluating our various priorities and satisfying our various stakeholders. To return to the beginning of this chapter, we cannot have forgotten the third mysterious stakeholder, which stands beside nature and the human part of nature, the name of which we were not then ready to discover. Let me try to name it now. It is the inner destiny of a given piece of landscape in itself: the suggestiveness of its beauty, its mysterious promise and potential for the future, the spiritual identity that the poet Gerard Manley Hopkins called its "inscape" (Abbott 1935: 66).

Such definitions may not help us much as scientists or practical landscape engineers. Luckily, however, there is a growing body of scientific theory and technological application that may take us a good deal of the way toward a conception of reclamational value that is both economical and rich. To rephrase the earlier questions: Can a better understanding of the nature of order, disorder, entropy, information, control, nonlinear dynamical processes, autonomy, and freedom help us develop a practical aesthetics and ethics of reclamation?

Let us take order and disorder first. What we see in nature is a hierarchy of orders, from the harmonic entanglement of the quantum world, with its maddening logic of superposition, through the reversible and deterministic order of classical dynamics, the irreversible and probabilistic order of thermodynamics and the self-transforming and complexifying order of evolutionary biology, to the promise-keeping, the deliberately law-abiding order of free human beings, creating new information at exponential rates of acceleration. Claude Shannon and John von Neumann have shown us that the increase of disorder, or entropic decay, in the world of thermodynamics can be the increase of informational order and organized complexity in a living organism, or even in a computer, programmed by human information creators (Shannon and Weaver 1949; von Neumann 1963: 341–42).[2] What is decay for the mature tree of one generation in the rain forest becomes the nurse log for the complex ecosystem of fungi, mosses, insects, small animals, and tree seedlings of the next.

But the terms "order" and "disorder" are becoming increasingly inadequate to describe the richness of the living world, or a fortiori the richness of the world of sentience and culture.

Until recently, the best we could do, with the available intellectual tools, in cobbling together some kind of reasonable account of the universe, and of our freedom in it, has been to devise some kind of combination between order and randomness, linear determinism and disordered noise. The title of Jacques Monod's book on biological evolution, *Chance and Necessity*, puts it well (1971). Perhaps we could describe both the emergence of new species and the originality and freedom of the human brain as a combination of random mutations and relatively deterministic selection.

But even then, there must be deep and subtle theoretical objections. Evolution seems to proceed in sudden jumps, not gradually; a new species does not emerge slowly, but rather leaps into being, as if drawn by a premonition of its eventual stable form. Another objection: without the right suite of species, the ecological niche does not exist; but without the ecological niche, the species does not. How do new niches emerge? Again, from a purely intuitive point of view, even four billion years does not seem nearly enough to produce the staggering variety and originality of form found among living species – birds of paradise, and slime molds, and hermaphroditic parasitical orchids, and sperm whales, and all. Most disturbing of all, it has become clear that the process of development, by which a fertilized egg or seed multiplies and diversifies itself into all the cells in all the correct positions necessary for an adult body, is not a mere following of genetic instructions embedded in the DNA blueprint. Rather, it is an original and creative process in itself, which produces a unique individual out of a dynamic and open-ended interplay of cells. The miracle is that the interplay could produce something in the end remotely resembling its twin siblings, not to mention its parents. It is as if the individual organism were drawn toward a beckoning form, and the genes were not so much blueprints, specifying that form, as gates, permitting the developmental process to rush toward its conclusion.

The same kinds of problems arise if we try to apply the chance-and-necessity model to the working of the human brain. Maybe "nature" and "nurture" do not exhaust the inputs. Can it make sense to speak of internal inputs, or forms that draw an appropriately prepared human brain into a specific competence, such as language? There seem to be a huge mass of internal, newly emergent laws and principles in such systems as ecologies, genomes, proteomes, natural languages, markets, political cultures, and so on that we have hardly begun to understand. And where did they come from, all of a sudden?

The dualism of order and disorder has been coming under increasing strain. Suppose we were to try to specify what an escape from this predicament might look like. We would have to distinguish between two kinds of order: a repressive, deterministic kind, and some other kind that does not have these disadvantages. We would also have to distinguish between two kinds of chaos: one that is simply random, null, and unintelligible, and another that can bear the seeds of creativity and freedom. If we were really lucky, the second kind of order might turn out not to be the antithesis of the second kind of chaos; the two might even be able to coexist in the same universe; best of all, they might even be the same thing! The extraordinary discovery that has been made – an astonishing stroke of good luck, a token of hope for the future – is that there really does seem to be the second kind of order, the second kind of chaos. And the two do seem to be the same thing.

This new kind of order, or chaos, or "chaorder," as some have called it, seems to be at the heart of an extraordinary range of interesting problems that have appeared as philosophers, mathematicians, scientists, and cybernetic technologists have tried to squeeze the last drops of the imponderable out of their disciplines. They include the biology and brain problems already mentioned; the problem of how to describe catastrophic changes and singularities by means of a continuous mathematics; the problem of how to predict the future states of positive-feedback processes; Gödel's paradox, which detaches the true from the provable; the description of phase changes in crystallography and electrochemistry; the phenomenon of turbulence; the dynamics of open systems and nonlinear processes; the observer problem in a variety of disciplines; the failure of sociological and economic predictive models to account for the rational expectations and second-guessing of real human subjects; the theoretical limitations of Turing machines, which, in certain circumstances, cannot turn themselves off (Penrose 2002: 40–86); the question of how to fit the fractal geometry of Benoit Mandelbrot into orthodox mathematics (1983: *passim*); the classification of quasicrystals and Penrose tilings; the issues of self-reflection, bootstrapping, and positive feedback in general; and, most troubling of all, the question of the nature of time.

In choosing the term "chaos" to describe this new imaginative and intellectual arena, its discoverers pulled off something of a public-relations coup without perhaps fully intending to. They could have called it "antichaos," which would have been just as accurate a term – in fact, a better one, since its implied double negative, "not-disorder," suggests something of its iterative depth. Maybe we could stick with "chaorder," which spans the region between deterministic chaos and constrained randomness. In order to understand the deeply liberating point of chaorder, we will need to go into the differences between deterministic linear order and chaotic emergent order, and between mere randomness and creative chaos.

Let us begin by considering an odd little thought experiment. Suppose we were trying to arrange a sonnet of Shakespeare in the most thermodynamically ordered way, with the least entropy. We cannot, for the sake of argument, break up the words into letters or the letters into line segments. The first thing we would do – which is the only sort of thing a strict thermodynamicist could do – is write the words out in alphabetical order: "a compare day I Shall summer's thee to?" As far as thermodynamics is concerned, such an arrangement would be more ordered than the arrangement "Shall I compare thee to a summer's day?," as composed by Shakespeare. Here, in a capsule, is the difference between deterministic linear order and chaotic emergent order.

We could even test the thermodynamic order of the first arrangement by a further experiment. Suppose we coded the words in terms of gas molecules, arranged in a row, the hottest ones corresponding to the beginning of the alphabet, the coldest ones to the end, and so on in alphabetical order. If left to themselves in a closed vessel, the molecules would, because of the increase in entropy over time, rearrange themselves into random alphabetical order (the hot and cold would get evenly mixed). Just as in a steam engine, where the energy gradient between hot steam and cold air can be used to do work, the movement of molecules can be employed to perform some (very tiny) mechanical task. And, as the second law of thermodynamics states, it would take somewhat more energy than we got out of it to put the molecules back into alphabetical order.

As arranged in sonnet 18, those words are already in more or less "random" alphabetical order. Yet most human beings would rightly assert that the sonnet order is infinitely more ordered than the thermodynamic, linear, alphabetical one. And in other respects the poem does seem to exhibit the characteristics of order. It could, if damaged by being rearranged, be almost perfectly reconstituted by a person who knew Shakespeare's work well. The sonnet can "do work": it has deeply influenced human culture and has helped to transform the lives of many students and lovers. It is an active force in the world precisely because it does not have the low-entropy simplicity of the alphabetical order that might enable it to do mechanical work. Here lies the basic distinction between "power" in the mechanical, political sense, and "power" in the sense of that mysterious creative influence of art.

Though we have distinguished between the two kinds of order, it is equally necessary to distinguish between the two kinds of chaos. One way of doing this is to differentiate between "white noise" and "flicker noise" (Davidsen and Paczuski 2002). White noise is made up of random amounts of energy at all frequencies. One could certainly imagine that one were listening to the sea when one heard acoustic white noise; there are even devices that make white noise in order to soothe people to sleep. But there is nothing in white noise to understand or interpret. On the other hand, flicker noise, which does not at first sound any different than white noise, is the "sound" made by a system that is ordered in itself and at the same time highly unstable and going through continuous internal adjustments by means of feedback. A good example of flicker noise is the sound made by a pile of sand onto whose apex new grains of sand are dropping one by one. There are many one-grain avalanches, fewer multigrain avalanches, fewer still mass avalanches, and only the occasional collapse of a whole slope. The sequence of avalanches obeys laws and forms an elegant fractal pattern when plotted on a graph. The sea itself makes a noise that is a fractal combination of interacting wave fronts, and provides another example of flicker noise.

The term that physicists use to talk about such feedback regimes is "nonlinear dynamical systems." Nature's most marvelously economical way of making exquisitely organized order is by the iterative action of an entity upon itself, or, even more effectively, the mutually iterative action of entities upon one another in a system – in fact, this mutual action is what constitutes a system as a system. We are familiar with the delicate fronds and whirlpools and paisleys that are produced simply by the iteration of the Mandelbrot algorithm on the screen of a computer, and with the astonishing richness of the pseudo-biological world that is generated by the simple rules of Conway's Game (Conway 2001: *passim*). In the world of nature, this wonderful economy of design genius is put to work everywhere.

What one hears when one hears flicker noise is the combination of events that are complexly but causally related. One is hearing the very process of emergent order arising from a nonlinear interactive dynamical process. But the examples of sand and sea are perhaps rather abstract. Flicker noise is not just the "sound" made by piles of sand and ocean surf. It is also what we get when we "listen" carefully to highly complex organic systems with strange attractors. The human brain, with remarkable consistency, finds such processes beautiful when it is able to disentangle their elements.

In the realm of artistic value, the idea of nonlinear systems generating emergent forms of order can prove very illuminating. When, in the move away from traditional societies to the modern state, we abandoned the old, value-laden notions of beauty, virtue, honor, truth, salvation, the soul, the divine, and so on, we suffered a genuine loss. Perhaps now we can re-found some of those beautiful notions upon a new-old basis. The strange attractor of a chaotic system can look very like an Ideal Form: although any instance of the outcome of such a system at work is only partial and apparently random, when we see all instances of the system's outcomes, we begin to make out a beautiful, if incomplete and fuzzy, shape. Might not virtues, ethics, values, and even, in a way, spiritual beings, be like those deep and beautiful attractors? And might there not be larger systems still, including many brains and the interactions of all of nature? Even the idea of freedom naturally emerges when we look at such systems. Ilya Prigogine shows us that self-organization can emerge out of open, entropy-increasing systems in a way that is radically irreversible and unpredictable; in other words, there exists no set of spaces in which a complex dynamical system's possible outcomes can be mapped (Prigogine 1997: 57–128).

Applying these thoughts on chaorder to reclamation, we might now be able to suggest that our goal should not simply be order, but the highest kinds of order. Or, rather, the combination of all the kinds of order that best serve the highest kinds of order – for the hierarchy of nature implies that a higher order unsupported by a full suite of the lower kinds of order withers quickly away. Designing according to the classical dynamics of a stable, graded tailings slope is not enough; we need to add in the thermodynamics of its continued erosion and sculpting by streams and its own chemical evolution, the biological succession and evolution of its future ecosystem, and the cultural potential of its location as an element in a hiking trail, a botanical garden, a fishing hole, a bike switchback, or a residential subdivision with a fine view that also provides a good place to raise children.

We can apply some of the same logic to the relationship between a reclaimed landscape and its local and global context: we should attempt to maximize the highest kind of order overall, while preserving as much local integrity and lower order as is consistent with the first goal. Though the bare rock and ice of a mountaintop might, by expensive technology, be made to support a mixed forest, the cost to the context of watersheds, scenery, and energy usage might be too great.

How do we identify the highest kinds of order? We have already posited the need for a perspective that includes change and evolution and growth. Mere change itself, however, can as well produce a barren and poisonous world, like those of most of the other planets in our solar system, as one with a rich suite of orders and complexities and information systems. Control systems are needed to guide change, so that it enriches and harmonizes rather than impoverishes and destroys. We are now beginning to find out which kinds of control systems are best at enriching and harmonizing. We know from classical thermodynamics that to enforce a desired result without consulting the existing flows and balances of the system in question can only result in a massive increase of disorder within the system, outside it, or both. We could perhaps make diamonds by crushing a forest, but the expense of energy and the destruction of the trees would be costly. There are wiser systems of control, like the surfer's use of a wave, that do not damage the order of the world, even as they add a new order to it (in the surfer's case, the aesthetic order of the ride). Good production design in the chemical industry now increasingly uses catalytic processes – natural cyclic energy feedbacks – to make their products. All biological systems are marvelously efficient at piggybacking on the energy and entropy and enthalpy that flow freely, abundantly, and on their own through the universe.

The key to the difference between enforced, coerced control, which is wasteful, and the persuasive kinds of control that can produce evolution, growth, and an increase of organized information, is feedback. The wise controller listens to and responds to the situation, like a lover rather than a tyrant, and works his or her will by recruiting the will and inclinations of the beloved. Thus, that "highest kind of order" that we are looking for is one in which there already exist systems of control; in which those systems are based on feedback rather than coercion or force; and in which the existing natural controllers welcome humans into the control room when they decide to enter. The control that the Chicxulub meteor exerted over the Earth's ecosystem was one of force and coercion. It was a natural case of environmental violation. The fact that it enabled some species to achieve even higher levels of encephalization and social organization is a tribute to the recuperative powers of life, not to the value of force. (But perhaps, as war is in human culture, force is, in nature, sometimes necessary.) The best form of government, to put the concept in terms of political philosophy, operates with the consent of the governed; as such, wise control applies to natural ecosystems as well as to nations.

Henry David Thoreau puts it well:

> Few phenomena gave me more delight than to observe the forms which thawing sand and clay assume in flowing down the sides of a deep cut on the railroad through which I passed

on my way to the village, a phenomenon not very common on so large a scale, though the number of freshly exposed banks of the right material must have been greatly multiplied since railroads were invented. The material was sand of every degree of fineness and of various rich colors, commonly mixed with a little clay. When the frost comes out in the spring, and even in a thawing day in the winter, the sand begins to flow down the slopes like lava, sometimes bursting out through the snow and overflowing it where no sand was to be seen before. Innumerable little streams overlap and interlace one with another, exhibiting a sort of hybrid product, which obeys half way the law of currents, and half way that of vegetation. As it flows it takes the forms of sappy leaves or vines, making heaps of pulpy sprays a foot or more in depth, and resembling, as you look down on them, the laciniated, lobed, and imbricated thalluses of some lichens; or you are reminded of coral, of leopard's paws or birds' feet, of brains or lungs or bowels, and excrements of all kinds. It is a truly grotesque vegetation, whose forms and color we see imitated in bronze, a sort of architectural foliage more ancient and typical than acanthus, chiccory, ivy, vine, or any vegetable leaves; destined perhaps, under some circumstances, to become a puzzle to future geologists. The whole cut impressed me as if it were a cave with its stalactites laid open to the light. The various shades of the sand are singularly rich and agreeable, embracing the different iron colors, brown, gray, yellowish, and reddish. When the flowing mass reaches the drain at the foot of the bank it spreads out flatter into strands, the separate streams losing their semicylindrical form and gradually becoming more flat and broad, running together as they are more moist, till they form an almost flat sand, still variously and beautifully shaded, but in which you can trace the original forms of vegetation; till at length, in the water itself, they are converted into banks, like those formed off the mouths of rivers, and the forms of vegetation are lost in the ripple-marks on the bottom.

The whole bank, which is from twenty to forty feet high, is sometimes overlaid with a mass of this kind of foliage, or sandy rupture, for a quarter of a mile on one or both sides, the produce of one spring day. What makes this sand foliage remarkable is its springing into existence thus suddenly. When I see on the one side the inert bank – for the sun acts on one side first – and on the other this luxuriant foliage, the creation of an hour, I am affected as if in a peculiar sense I stood in the laboratory of the Artist who made the world and me – had come to where he was still at work, sporting on this bank, and with excess of energy strewing his fresh designs about. I feel as if I were nearer to the vitals of the globe, for this sandy overflow is something such a foliaceous mass as the vitals of the animal body. You find thus in the very sands an anticipation of the vegetable leaf. No wonder that the earth expresses itself outwardly in leaves, it so labors with the idea inwardly. The atoms have already learned this law, and are pregnant by it. The overhanging leaf sees here its prototype. Internally, whether in the globe or animal body, it is a moist thick lobe, a word especially applicable to the liver and lungs and the leaves of fat (leibo, labor, lapsus, to flow or slip downward, a lapsing; lobos, globus, lobe, globe; also lap, flap, and many other words); externally a dry thin leaf, even as the *f* and *v* are a pressed and dried *b*. The radicals of lobe are *lb*, the soft mass of the *b* (single-lobed, or *B*, double-lobed), with the liquid *l* behind it pressing it forward. In globe, *glb*, the guttural *g* adds to the meaning the capacity of the throat. The feathers and wings of birds are still drier and thinner leaves. Thus, also, you pass from the lumpish grub in the earth to the airy and fluttering butterfly. The very globe continually transcends and translates itself, and becomes winged in its orbit. Even ice begins with delicate crystal leaves, as if it had flowed into moulds which the fronds of waterplants have impressed on the watery mirror. The whole tree itself is but one leaf, and rivers are still vaster leaves whose pulp is intervening earth, and towns and cities are the ova of insects in their axils.

(1969: 327–28)

Let us look at two elements of this magnificent piece of prose. The first is that Thoreau does not draw any line between the creative dynamical processes of nature and the creative dynamical processes of human culture. The place, after all, is a railroad cutting – a pretty large local disturbance of the existing ecosystem by what was then the most recent and futuristic technology, the railroad. Nature, Thoreau realizes, is not averse to disturbance and often reveals its finest pattern making in response. We know now, likewise, that forests and prairies need interruptions and scars to be able to support their rarest and most spectacular fauna and flora – life flourishes at edges and margins, even better than in heartlands and refuges. Human cities are part of the pattern, from which even human language emerges.

And here we come to the second element I want us to observe in Thoreau's passage. All of that play on language, on the etymologies of human words, celebrates the emergence, from the lobed and branchy human body, of the lobed and branchy family tree of languages and the lobed and branchy process of human metaphor making. The sandy bank, no less than the free human mind, is at liberty to flow out into a great variety of possible outcomes. For Thoreau, time is continually bifurcating, branchy, whether in the decisions of persons or in the next action of a dynamical physical process, when the feedbacks are so complex that every outcome is a butterfly effect – sensitively dependent on initial conditions – and initial conditions are unrecoverable from the result.

Let us look at this from a more philosophically exact point of view. A single cause can have multiple effects; the laws of nature (which are themselves the historical result of earlier cause–effect events) constrain those possible effects but do not reduce them to one. This is what I mean by "branchiness." All the effects that do not completely neutralize one another take place, distributed in number and intensity according to their probability, along a bell-shaped curve. The difference between the number of bytes required to describe the "cause state" and the number required to describe the later "effects state" is the measure not only of the increase in its entropy and in its thermodynamic disorder, but also in the amount of new information that has entered the world. That difference also describes the degrees of freedom of the initial state (if there is only one "branch," a single possible effect, the system has no freedom; if there are many "branches," the system has free play to that extent). However, an event can bring about a set of effects that are either mutually reinforcing or dissonant; the mutual reinforcement of those effects is the measure of its informational (as opposed to its thermodynamic) order. If that state of mutual reinforcement limits the number of new effects, we may call it a "barren order"; if it makes possible a further set of mutually reinforcing but productive effects, we may call it a "rich order." The biological definition of evolutionary-reproductive success is very similar: as many offspring as is consistent with the survival-to-reproduce of the offspring's offspring, and of that offspring's offspring in turn.

Thus, we have a theoretical basis for choosing one set of actions in the environment over another, that is, a basis for deciding what is or is not good reclamation. Good reclamation is reclamation that creates the greatest freedom consistent with the richest order in the effect-state it produces.

This very dry and theoretical formulation can be given life and color by the reflection that, as human animals, we already possess a remarkably sensitive and intuitive sensory and cognitive system for detecting free and informational-order-rich systems: it is our aesthetic sense, as demonstrated marvelously by the quotation from Thoreau.

Notes

1 See Berger's *Reclaiming the American West* (2002).
2 See also Bennett (1982: 905–40) and Bennett and Landauer (1985: 48–56).

Bibliography

Abbott, C.C. (ed.) (1935) *The Correspondence of Gerard Manley Hopkins and Richard Watson Dixon*, vol. 1, London: Oxford University Press.

Bennett, C.H. (1982) 'The thermodynamics of computation – a review,' *International Journal of Theoretical Physics* 21: 905–40.

Bennett, C.H. and Landauer, R. (1985) 'The fundamental physical limits of computation,' *Scientific American* 253, 1: 48–56.

Berger, A. (2002) *Reclaiming the American West*, New York: Princeton Architectural Press.

Conway, J.H. (2001) *On Numbers and Games*, Boston, MA: A.K. Peters.

Davidsen, J. and Paczuski, M. (2002) 'Of noise from correlations between avalanches in self-organized criticality,' *Physical Review* 66, 5, November: 50101–104.

Mandelbrot, B. (1977; rev. edn 1983) *The Fractal Geometry of Nature*, New York: W.H. Freeman.

Monod, J. (1971) *Chance and Necessity: An Essay on the Natural Philosophy of Modern Biology*, trans. A. Wainhouse, New York: Vintage.

Penrose, R. (2002) *The Emperor's New Mind*, New York: Oxford University Press.

Prigogine, I. (1997) *The End of Certainty*, New York: The Free Press.

Shannon, C. and Weaver, W. (1949) *The Mathematical Theory of Communication*, Urbana-Champaign, IL: University of Illinois Press.

Thoreau, H.D. (1969) *Walden*, ed. W. Thorp, Columbus, OH: Charles E. Merrill.

von Neumann, J. (1963) 'Probabilistic logics and the synthesis of reliable organisms from unreliable components,' in A.H. Taub (ed.) *John von Neumann's Collected Works*, vol. 5, New York: Macmillan.

Chapter 2
Disturbance ecology and symbiosis in mine-reclamation design

Peter Del Tredici

Nature is a process, not a product, and its primary driver is the Darwinian principle of natural selection, which is succinctly described as the survival of the fittest. There is no morality in this evolutionary process, only the reality of which individuals live and which ones die. Over the past two hundred years, human activities associated with industrialization have brought about changes in the global environment that are unprecedented in both scale and magnitude. This disruption of natural ecosystems has led to the breakup of long-established biological associations among organisms that were once native to an area and the creation of new niches that are open to a cosmopolitan array of non-native species. In the context of these changes, natural selection dictates greater reproductive success and ecological dominance for those species that are best adapted to the new environmental conditions, regardless of their past evolutionary or ecological history (Gould 1998: 2–10).

So the question facing today's landscape architects and planners is this, *what model should be used for rebuilding sites that have been heavily degraded by human activity, such as mine spoils or the post-industrial landscapes that form the core of many urban centers?* For brevity's sake, I have narrowed the options down to two: reclamation or restoration (Harris *et al.* 1996: 16–18).

Reclamation, which can also be referred to as revitalization, starts with the assumption that the ecological clock cannot be turned back to an earlier time. Its broad goals are to minimize the negative impacts that the site may have on the surrounding environment and to maximize its aesthetic and ecological functionality. Reclamation projects are usually large scale and heavily disturbed and cry out for some form of productive reuse. A core principle of reclamation is that everything that happens to a given piece of ground becomes an inseparable part of what it can become in the future.

Restoration, on the other hand, starts with the linked assumptions that it is both possible and desirable to reestablish some portion of the original ecological conditions of a site. People who advocate strict restoration face two very difficult questions: to what former time period should the site be restored? And how should one cope with the unpredictable environmental changes that impact the site?

Succession in the modern world

From an ecological perspective, the issues of reclamation and restoration ultimately revolve around the issue of *succession*, the term used to describe the change over time in the composition of the plants, animals, and microbes that inhabit a given area. Typically, two types of succession are recognized: primary, which occurs on bare ground with no past biological history – glaciation or volcanic eruptions, for example – and secondary, which occurs when organisms replace one another, as a result of changing landscape conditions, such as occurs following agriculture, logging, or fire. Prior to World War II, ecologists tended to view succession as an orderly process, leading to the establishment of a climax, or steady-state, community that, in the absence of disturbance, was capable of maintaining itself indefinitely. I refer to this as the "Disney" version of ecology: stable and predictable, with all organisms living in perfect harmony. In the 1950s, a younger generation of ecologists began to challenge this orthodox view, eventually formulating what is now known as the theory of "patch dynamics," embracing natural disturbances as an integral part of a variable and unpredictable succession process (Barbour 1995: 233–55). The key concept here is that the nature, timing, and intensity of the disturbances are critical factors – together with climate and soil – in determining the composition of successive generations of vegetation. Succession is seen as a stochastic process with an uncertain outcome, as opposed to one with a predetermined end point. From the modern ecological perspective, the apparent stability of current plant associations is an illusion; the only certainty is that things will be substantially different within thirty or forty years (Fisk and Niering 1999: 483–92).

When one broadens the traditional definition of disturbance to include the effects of acid rain on the Earth's surface and of carbon dioxide enrichment on its atmosphere, it becomes clear that there is no place on Earth that has not experienced some level of alteration as a result of human activities (Vitousek 1993: 1861–76). The absurd position that global warming has not yet been proven is based on the assumption that humans have the capacity to understand – at a detailed level – how the world's climate system actually works. Indeed, the scariest thing about climate change is the uncertainty of how it will play out on the ground. When and if scientists get around to predicting accurately the effects of pumping massive amounts of carbon dioxide and nitrous oxides into the atmosphere, it will be far too late to do anything about it (Houghton 2004: 216–39).

One particularly problematic aspect of the restoration concept is its denial of the inevitability of ecological change. Implicit in much of the popular writing on the subject is the assumption that the plant and animal communities that existed in North America prior to European settlement can be returned to some semblance of their original composition. The fact that the environmental conditions that led to the development of these pre-Columbian habitats no longer exist – and can never be re-created – does not seem to count for much. In my opinion, the support for such "faith-based" notions of restoration has more to do with the ethical values – of wanting to do the right thing – than with the ecological reality.

Historical experience in eastern New England clearly shows that, even when the individual components of former ecosystems make successful comebacks, they tend to function differently than they did in the past, because irreversible changes have occurred in other parts of the ecosystem (see Figure 2.1). This is best exemplified by herds of white-tailed deer (*Odocoileus virginianus*) that were formerly controlled by the hunting activities of woodland Indians but today roam the countryside in large herds, selectively browsing native species while ignoring the unpalatable invasive plants (Krech 1999: 151–74). In the process, they manage not only to annoy home owners but also to alter long-established patterns of forest succession (Foster *et al.* 2002: 1337–57). The dynamic nature of interactions among people, plants, animals, and introduced pathogens in today's world is producing novel ecological conditions with unpredictable consequences for the future (Gregg *et al.* 2003: 183–87).

Extreme disturbance leads to new ecological forms

The application of the concept of ecological restoration to the urban habitat is particularly problematic, given the abundance of storm-water runoff, road salt, heat build-up, air pollution, and soil compaction that characterizes metropolitan centers. Indeed, the critical question facing

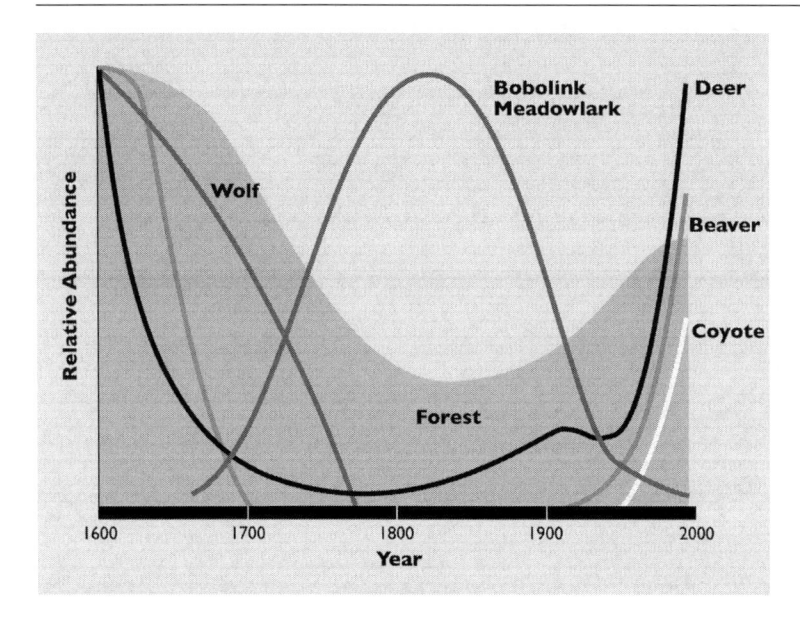

Figure 2.1 Changes in land-use, forest-cover, and animal populations in New England, from 1700 to 2000. Illustration courtesy of Harvard Forest, Petersham, Massachusetts.

landscape professionals who work in these areas is not *What plants grew in the past?* but *What will grow there in the future?* Starting in the early 1800s and continuing through the present, ornamental plants from around the world have been brought together in our cities and suburbs in order to enhance their livability (Reichard and White 2001: 103–13). Like it or not, a small percentage of these horticultural introductions have adapted well to their new homes and begun reproducing on their own. Regardless of the disparaging labels often applied to these "naturalized" species, many of them are actually performing significant ecological functions, including heat reduction, water and air filtration, mineral cycling, and carbon fixation and storage (De Wet *et al.* 1998: 237–62).

A good example of this is the common reed, *Phragmites australis*, which is native to Europe and central Asia, as well as to North America, where it grows in brackish wetlands up and down the East Coast, most dramatically in the meadowlands that border the New Jersey Turnpike west of Manhattan (see Figure 2.2). While *Phragmites* is often portrayed as the ultimate invasive species because it tends to crowd out other vegetation, it is actually mitigating pollution by absorbing a great deal of the nitrogen and phosphorous that accumulates in degraded wetlands. Indeed, in temperate ecosystems worldwide, the common reed is widely used for wetland phytoremediation (Meuleman *et al.* 2002: 712–21). From the functional perspective, the presence of *Phragmites* in this landscape can be viewed as a symptom of environmental degradation, rather than as its cause. It turns out that many invasive plants have a similar kind of Jekyll-and-Hyde impact on the local ecology, pushing out some native plants while providing food and shelter for a variety of native animals (D'Antonio and Meyerson 2002: 703–13; Thacker 2004: 182–87).

Regardless of one's feelings about the cosmopolitan assemblages of plants that now populate America's sprawling cities, they are in the process of becoming the urban forests, fields, and wetlands of tomorrow. In a very real sense, the diversity and spontaneity of these "immigrant" communities mirror that of modern human society. Indeed, the very processes that have led to the globalization of the world economy – unfettered trade and travel among nations – have resulted in the globalization of the environment (Normile 2004: 968–69).

Figure 2.2 The common reed, *Phragmites australis*, growing in the New Jersey meadowlands. Photograph by Peter Del Tredici.

Although much of the preceding discussion is based on my personal experience of working in cities, the underlying principles apply to other drastically disturbed sites, such as post-industrial and post-mining landscapes. In the latter case, all existing vegetation, as well as the topsoil, has been removed, to expose coal-rich seams or mineral-rich rock, which is extracted in its entirety. This net loss of structural material creates voids, or tailing ponds, which are often filled with waste rock and a highly acidic cocktail of minerals that can be toxic to both animals and plants. In short, mining results in the total destruction of existing biological communities and the creation of geological conditions reminiscent of a much earlier successional state. Any concept of restoration of the land to its pre-mining state is beyond the realm of possibility.

Four steps to ecologically sound mine reclamation

So what can landscape professionals, working with mine wastes on the ground, do to cope with the widespread environmental devastation and ecological uncertainty that are integral to the modern mining process? The first step, of course, is to make sure that the substrate can support the growth of plants. This means coming to grips with the chemical and physical properties of mine spoils that inhibit the growth of plants, including pHs that are either too high or too low, the abundance of toxic minerals, and the coarse, rocky texture that severely limits water-holding capacity. Suffice it to say that, without adequate remediation of these basic substrate problems, there is no hope of ever getting anything to grow. Essentially one is dealing with a situation that is analogous to primary succession, in which the soil ecosystem has to be built up from scratch (Hutchings 2002: 359–76).

The second step of this re-vegetation strategy is closely tied to the first: to spare no effort in enriching degraded land with organic matter in the form of cover crops or mulch. Not only does organic matter jump-start the soil-forming process by increasing water-holding capacity, it also facilitates nutrient cycling by promoting the growth of beneficial microorganisms such as symbiotic mycorrhizal fungi and nitrogen-fixing bacteria (see Figure 2.3). At impoverished mine sites, where growing conditions are extremely stressful because of high temperatures, drought, low fertility, and pH-related toxicity, a plant's symbiotic relationships with soil microbes make the difference between life and death.

In 1966, one of the pioneers of reclamation biology, J.R. Shramm, demonstrated the important role that *ectomycorrhizal* fungi (ECM) play in allowing a variety of woody plants to spontaneously colonize anthracite-mining wastes in Pennsylvania (Shramm 1966: 131–41). One of the species that he identified, *Pisolithus tinctorius*, is now commercially available under the name "Pt" and is widely used as an inoculant for tree seedlings in reclamation projects throughout the world. By means of their extensive underground mycelial networks, ECM facilitate the cycling of nitrogen and phosphorus between and among the various species of plants they are connected to (Simard *et al.* 1997: 579–82). It has been estimated that plant roots with attached ectomycorrhizal fungus have an absorptive surface area roughly a hundred times greater than that of roots without mycorrhizae (Smith and Reed 1997: 276–89).

In contrast to woody plants, herbaceous perennials and grasses typically develop symbiotic relationships with a different type of fungus, known as *endomycorrhizae*, or *vesicular-arbuscular mycorrhizae* (VAM). These fungi are distinguished from ECM by three main features: VAM do not change the external morphology of the host plant's root system, their microscopically thin hyphae penetrate the plant's root cells, and they never produce the above-ground fruiting

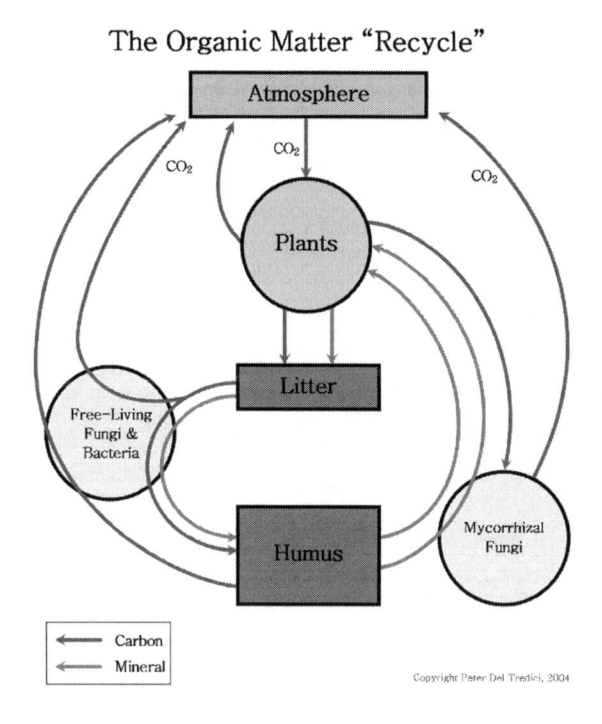

Figure 2.3 The Organic Matter "Recycle."

bodies (i.e. mushrooms) that are a familiar feature of forested habitats where ECM are dominant. The genus *Glomus* is among the most widely distributed of the VAM fungi, and it infects a wide variety of plants from corn to giant redwoods. It is relatively easy to grow in culture and is readily available from commercial sources. At impoverished mine-reclamation sites, both ECM and VAM are essential for the successful establishment of any vegetation, and their use should be specified either as an inoculant for contract-grown nursery stock or as a soil additive when waste-rock is direct-seeded (Smith and Reed 1997: 470–75). Recent research has documented the potentially significant role that mycorrhizae can play in transforming and detoxifying soils contaminated with heavy metals and hydrocarbon compounds (Gadd 2004: 60–70).

Among bacteria, species in the genus *Rhizobium* are noteworthy for their ability to form symbiotic relationships with many leguminous plants – such as the well-known black locust, *Robinia pseudoacacia*. The bacteria are localized in root nodules, where they "fix," or convert atmospheric nitrogen to a form that can be used by the plant, thereby allowing it to flourish in nutrient-poor soil. Many non-leguminous woody plants – including alder (*Alnus spp.*) and sweetfern (*Comptonia peregrina*) – have a similar nitrogen-fixing relationship with filamentous bacteria in the genus *Frankia*, which allows them to colonize both wet and dry, infertile sites (Callaham *et al.* 1978: 899–902; Del Tredici 1996: 26–31) (see Figures 2.4 and 2.5). Both of these nitrogen-fixing plant groups perform well on reclamation sites, and they should be an essential part of any reclamation project.

The third step in my re-vegetation strategy deals with the issue of plant selection. In this regard, I advocate *not limiting* planting designs to the palette of native species that historically grew on the site. Imposing such a limitation not only reduces the chances of getting plants established on the site, but also the aesthetic possibilities for the site in the future. I propose that sustainability be the standard for deciding what to plant on drastically disturbed sites. According

Figure 2.4 Red alder, *Alnus rubra*, growing naturally on a talus slope in Alaska. Photograph by Peter Del Tredici.

Disturbance ecology and symbiosis

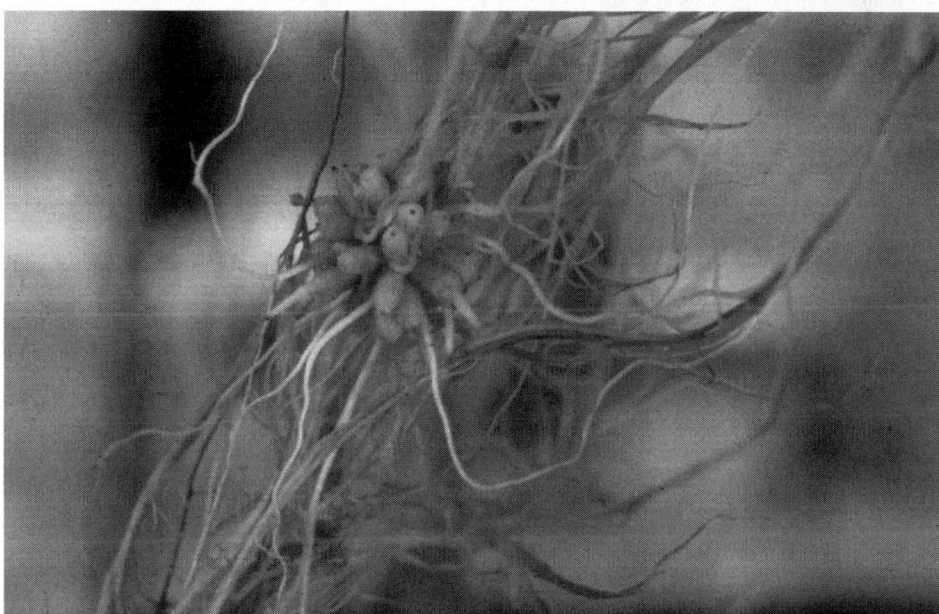

Figure 2.5 Top, the sweetfern, *Comptonia peregrina*, growing on a gravelly roadside bank in central Massachusetts; bottom, close-up view of a sweetfern root nodule growing in the laboratory. Photographs by Peter Del Tredici.

to my definition of this overused word, sustainable landscape plants are those that can tolerate the conditions that prevail on a given site, require minimal levels of maintenance in order to get established, and display a limited capacity to spread by seed into surrounding natural areas. In general, they are tolerant of a broad range of light, moisture, and nutrient conditions (Del Tredici 2001a: 10–18). Landscapes that are designed with such sustainable plants – including both native and introduced species – will not only be less costly to maintain over time, but also more adaptable to the unpredictable weather patterns and pathogen problems that clearly loom in the future. When it comes to selecting plants for re-vegetating disturbed sites, American designers have much to learn from their European counterparts, who have a long tradition of using cosmopolitan plant associations to create stable, naturalistic landscapes in a wide variety of habitats (Hitchmough and Dunnett 2004: 1–22).

For mining-reclamation projects, in particular, it is important to select trees and shrubs that have a strong capacity to produce new shoots either from their roots ("suckering"), as in the case of quaking aspen (*Populus tremuloides*), black locust (*Robinia pseudoacacia*), and gray dogwood (*Cornus racemosa*) (see Figure 2.6), or from the base of their trunks following traumatic injury ("coppicing"), as in the case of sweet birch (*Betula lenta*) (see Figure 2.7). In the former instance, the species are able to spread over time and cover large areas; in the latter, they are able to persist at stressful sites by resprouting after damage from predators or severe weather (Del Tredici 2001b: 121–40). Such disturbance-adapted, "pioneer" species are clearly the best choice for re-vegetating steep mining sites, in which the succession process starts with bare rock. In Table 2.1, I have summarized the life-history traits that, in general, preadapt woody plants for re-vegetating mine reclamation sites.

The fourth and final aspect of my strategic approach to reclamation requires an acknowledgment – early on in the design process – of the need for ongoing maintenance at all constructed landscapes, regardless of scale. All too often the concept of sustainability is misinterpreted to mean self-sustaining, a fantasy that is as false in ecology as it is in horticulture. Standard landscape maintenance practices – including irrigation, weeding, mulching, and replanting – are necessary to promote the successful establishment of any new planting, regardless of the theories it is based on. From the horticultural perspective, a truly sustainable landscape design is one that is in balance with the financial resources available to maintain it (Koningen 2004: 256–92).

Disturbance ecology and symbiosis

Figure 2.6 Opposite, all of the stems in this clonal grove of quaking aspen, *Populus tremuloides*, in the Rocky Mountains of Colorado arise from a common root system; top, the black locust, *Robinia pseudoacacia*, reproducing from root suckers on a vacant lot in Boston; bottom, the gray dogwood, *Cornus racemosa*, spreading by root suckers along an abandoned roadway in Boston. Photographs by Peter Del Tredici.

Figure 2.7 The black birch, *Betula lenta*, growing on a hundred-year-old pile of weathered slate slag, in the town of Harvard, Massachusetts. The tree displays a multi-stemmed growth form.
Photograph by Peter Del Tredici.

Table 2.1 A summary of the basic life-history traits that preadapt some woody plants for growth on mine reclamation sites.

- germinate readily from seed
- be relatively fast growing
- be tolerant of extreme sun and wind exposure as well as high soil and air temperatures
- possess a strong capacity for vegetative regeneration from suckers, stump sprouts, rhizomes, or branch layers
- be able to grow in soils with high concentrations of heavy metals
- be able to grow in soils with low pHs
- be tolerant of drought induced by coarse textured or highly compacted soils
- on wet sites, be tolerant of saturated soils with low oxygen tensions and high concentrations of toxic compounds
- be able to form symbiotic relationships with a broad range of both ecto- and endomycorrhizae
- on low nutrient sites, be able to form symbiotic relationships with nitrogen-fixing bacteria

What would Olmsted do?

The ecological approach to mine-reclamation design that is outlined above requires changing the focus of the design process from its traditional emphasis on form (i.e. the species list) to an emphasis on ecosystem function (i.e. energy flows, water use, mineral cycling, and carbon sequestration). In practical terms, this translates into installing more, smaller, bare-root plants, as opposed to fewer, larger, ball-and-burlap or containerized plants, thereby allowing the microclimatic features of the site to determine which individuals live and which die. The designer's role is to develop the palette of plants to be used on the site and to determine the placement of those plants in relation to gradients of topography, aspect, soil type, and moisture.

The technique of over-planting small material requires abandoning the age-old practice of assigning individual plants to fixed locations in a planting plan and replacing it with mixed plantings of trees, shrubs, and herbaceous perennials. Adopting this change in planting strategy is difficult for some landscape architects, because it means giving up a measure of control over the planting design to the plants themselves. But one can take heart in the fact that the great Frederick Law Olmsted always over-planted his public parks, with the intention that they would later be thinned out. In his plan for the park at Niagara Falls, written in 1889, he articulated a philosophy that is directly applicable to today's large-scale reclamation projects:

> They [the plantings] are to be thinned out gradually as they come to interlock, until, at length, not more than one-third of the original number will remain, and these, because the less promising will have constantly been selected for removal with little regard to evenness of spacing, will be those of the most vigorous constitution, those with the greatest capabilities of growth, and those with the greatest power of resistance to attacks of storms, ice, disease and vermin. Individual tree beauty is to be but little regarded, but all consideration to be given to beauty and effectiveness of groups, passages, and masses of foliage.
> (Olmsted 1928: 370)

Olmsted's instructions make it clear that he viewed landscape design as a dynamic rather than a static process. Sustainability, if the term has any meaning at all, is about form following function and about allowing landscapes the time they need to develop complexity and character through long-term interactions with their environment. It is not about moving big trees around like chess pieces to create instant, climax-forest effects. Landscape architects can embrace this concept of sustainability by making ecological functionality and environmental adaptability primary design goals.

Bibliography

Barbour, M.G. (1995) 'Ecological fragmentation in the fifties,' in W. Cronin (ed.) *Uncommon Ground*, New York: Norton.

Callaham, D., Del Tredici, P., and Torrey, J.G. (1978) 'Isolation and cultivation *in vitro* of the actinomycete causing root nodulation in *Comptonia*,' *Science* 199: 899–902.

D'Antonio, C. and Meyerson, L.A. (2002) 'Exotic plant species as problems and solutions in ecological restoration: a synthesis,' *Restoration Ecology* 10: 703–13.

Del Tredici, P. (1996) 'A nitrogen fixation: the story of the *Frankia* symbiosis,' *Arnoldia* 55, 4: 26–31.

—— (2001a) 'Survival of the most adaptable', *Arnoldia* 60, 4: 10–18.

—— (2001b) 'Sprouting in temperate trees: a morphological and ecological review,' *Botanical Review* 67, 2: 121–40.

De Wet, A.P., Richardson, J., and Olympia, C. (1998) 'Interactions of land-use history and current ecology in a recovering "urban wildland,"' *Urban Ecosystems* 2: 237–62.

Fisk, J. and Niering, W.A. (1999) 'Four decades of old field vegetation development and the role of *Celastrus orbiculatus* in the northeastern United States,' *Journal of Vegetation Science* 10: 483–92.

Foster, D., Motzkin, G., Bernardos, D., and Cardoza, J. (2002) 'Wildlife dynamics in the changing New England landscape,' *Journal of Biogeography* 29: 1337–57.

Gadd, G.M. (2004) 'Mycotransformation of organic and inorganic substrate,' *Mycologist* 18: 60–70.

Gould, S.J. (1998) 'An evolutionary perspective on strengths, fallacies, and confusions in the concept of native plants,' *Arnoldia* 58, 1: 2–10.

Gregg, J.W., Jones, C.G., and Dawson, T.E. (2003) 'Urbanization effects on tree growth in the vicinity of New York City,' *Nature* 424: 183–87.

Harris, J.A., Birch, P., and Palmer, J.P. (1996) *Land Restoration and Reclamation: Principles and Practice*, Essex: Addison Wesley Longman.

Hitchmough, J. and Dunnett, N. (2004) 'Introduction to naturalistic planting in urban landscapes,' in N. Dunnett and J. Hitchmough (eds) *The Dynamic Landscape*, London: Spon Press.

Houghton, J. (2004) *Global Warming: The Complete Briefing* (third edn), Cambridge: Cambridge University Press.

Hutchings, T.R. (2002) 'The establishment of trees on contaminated land,' *Arboricultural Journal* 26: 359–76.

Koningen, H. (2004) 'Creative management,' in N. Dunnett and J. Hitchmough (eds) *The Dynamic Landscape*, London: Spon Press.

Krech, S. (1999) *The Ecological Indian: Myth and History*, New York: Norton.

Meuleman, A.F.M., Beekman, H.P., and Verhoeven, J.T.A. (2002) 'Nutrient retention and nutrient-use efficiency in *Phragmites australis* stands after wastewater application,' *Wetlands* 22: 712–21.

Molina, R.J. (1984) 'Commercial vegetative inoculum of *Pisolithus tinctorium* and inoculation techniques for development of ectomycorrhizae on bareroot tree seedlings,' *Forest Science Monograph* 25: 1–101.

Normile, D. (2004) 'Expanding trade with China creates ecological backlash,' *Science* 306: 968–69.

Olmsted, F.L. (1928) 'Observations on the treatment of public plantations, more especially relating to the use of the axe,' in *Forty Years of Landscape Architecture, the Professional Papers of Frederick Law Olmsted, Senior: Central Park* (vol. 2), New York: Putnam.

Reichard, S.H. and White, P.S. (2001) 'Horticulture as a pathway of invasive plant introductions in the United States,' *BioScience* 51: 103–13.

Shramm, J.R. (1966) 'Plant colonization studies on black wastes from anthracite mining in Pennsylvania,' *Transactions of the American Philosophical Society*, new series, 56, part 1: 1–194.

Simard, S.W. *et al.* (1997) 'Net transfer of carbon between ectomycorrhizal tree species in the field,' *Nature* 388: 579–82.
Smith, S.E. and Reed, D.J. (1997) *Mycorrhizal Symbiosis*, San Diego, CA: Academic Press.
Thacker, P.D. (2004) 'California butterflies: at home with aliens,' *BioScience* 54: 182–87.
Vitousek, P. (1993) 'Beyond global warming: ecology and global change,' *Ecology* 75: 1861–76.

Chapter 3
Gold and the gift:
theory and design in a mine-reclamation project

Rod Barnett

Introduction

This chapter outlines a theoretical agenda for the reclamation of a landscape radically altered by gold mining. The purpose of the chapter is to demonstrate that even quite modest alterations of disturbed terrain can open up the rich theoretical seams that weave through the concept of waste.

The many disciplines involved in landscape reclamation are linked by a web of connective intellectual tissue – nonlinear dynamics – that can be used to develop meaningful design strategies for waste landscapes. The proposition discussed here links philosophically both to this network of disciplines (ecology, archaeology, landscape architecture, economics) and to the larger issues of cultural value, atonement, and poetics that gather around the idea of reclamation. It is an attempt to combine the problems of how to do unto others (such as miners, tussock plants, and lizards) with provision for the enhanced human experience of the world we share with these others. Something, in other words, for everybody. In this piece, I draw on three sets of theoretical constructs – those of Ilya Prigogine's nonlinear systems, of Georges Bataille's "gift," and that of historical pleasure gardens – to develop a proposition for the design of a retired mine site in the tussock country of Otago in New Zealand's South Island.

The Macraes gold site

Oceana Gold (NZ) Limited owns and operates the Macraes Gold Operation, situated ninety kilometers north of Dunedin, the provincial capital of Otago (see Figure 3.1). Oceana Gold is New Zealand's largest gold producer, and the production from Macraes accounts for approximately 50 percent of the country's annual gold take. It is considered good mining practice by Oceana Gold to rehabilitate its Macraes mine site to an appropriate land use after mining. This is also a requirement of the resource consents that the local governmental authorities – Waitaki District Council and Otago Regional Council – issued for the extraction process.

Gold was first discovered in the Macraes Flat area by alluvial prospectors in 1862. A flourishing canvas town was soon established, and in 1866, quartz-bearing lodes were found. Very little gold was won from the reefs, and, with the consequent waning of interest, reef-mining activity declined and ceased about 1868. Twenty years later, an upsurge in gold-mining interest led to the reopening of the Highway Reef near Round Hill and the discovery of several other reefs in the district. From 1890 to 1930, an estimated production of a hundred thousand tons of ore yielded approximately fifteen thousand ounces of gold. Gold mining stopped for WW II and recommenced as an open cast system in the mid-1980s.

Theory and design in mine reclamation

Figure 3.1 Location of Oceana Gold Mining site in the South Island of New Zealand.

There are numerous heritage mining features on the Macraes site, including hard-rock mines, batteries, a Chinese miner's mud-brick house, a Chinese miner's camp site, alluvial gold workings, nineteenth-century tailings, earth-dam remains, and early mining implements. Together these historic features tell the story, across a vast landscape, of how alluvial and hard-rock mining operated, giving a vivid impression of the origins of the gold-mining industry in New Zealand.

The context: Macraes Flat

The township of Macraes Flat is located within the vast reclamation site, on land that is now either owned or leased by Oceana Gold. It dates from the gold-mining days of the nineteenth century, and a number of early structures still exist: a bootmaker's shop, a grain store, and a billiards room, as well as a vicarage, a church, and a number of stone structures, such as a hotel, a house, and walls. As gold extraction became more difficult, farmers moved into the territory; and, for much of the twentieth century, Macraes Flat served a community of two or three hundred families whose livelihood was gained from raising cattle and sheep. The town is regarded by Oceana Gold as an integral part of the reclamation strategy, and it is one of the trickier elements of negotiating the proposal, because the people who live there – miners and farmers – have an ambivalent attitude toward the mining company. They are people who have labored long and hard, as their forebears did, to win a living from an unforgiving, intractable landscape. For reasons to do with this very issue – and with intrinsic qualities of the theoretical proposition – this chapter focuses on landscape interventions designed for the township itself. These interventions are low-key, subtle rearrangements of historical terrain, and they are intended to serve the purpose of permitting a reconnection with something larger than a working life at the same time that they validate and complicate the human condition – that is, the necessity of labor.

Before these interventions are considered, however, it is important to contextualize them within the design of the larger, decommissioned mining site. The rehabilitation plan for the Macraes Gold Operation provides for the development of a Heritage and Art Park. It is to be a recreational resource that combines history, art, designed landscapes, the natural environment, ecology, technology, and traditional and alternative agriculture into a self-sustaining destination for tourism purposes. All of the stakeholders in the property (Oceana Gold, the Otago Regional Council, and the local population) expect this singular blend of tourism, conservation, and arts practice to encourage future growth and development for the community once mining has ceased.

The key aspects of the proposal are, first, ecological reconstruction; second, infrastructural provision; third, the preservation of historical features; and, fourth, the creation of landscape interventions in the form of gardens and significant artworks. The extent of the Heritage and Art Park includes the mine site itself (the open pits, waste-rock stacks, tailings, and dams), the township, and the surrounding farmland, comprising pasture, wetlands, and shelter belts. Beyond the farmed terrain, but within the park boundaries, there is a "wilderness" of native vegetation, of tussock lands, ravines, rocky slopes, and high plains (see Figure 3.2).

Figure 3.2 Aerial showing proximity of Macraes township to mining site.

Theory

There are two ways in which gold mining is inextricably bound up with the production of waste. Mining companies such as Oceana Gold produce waste – that which is not valued because it is not gold – as a by-product of their extraction operations; this is what Alan Berger calls "actual waste" (Berger 2006: 14). The landscape that mining has transformed from its previous state – be that "natural" or "cultural" – is turned into a waste landscape, a Bergerian "wasted place," that is, an abandoned or contaminated site. How is this overproduction, this production in excess of the requirements, to contribute to the notion of "rehabilitation" or, as Berger calls it, "reintegration" into "flexible and aesthetic design strategies" for "higher social, cultural and environmental benefits" (Berger 2006: 45)? Is it possible to make the redesign of a mined landscape into something more than an amenity?[1]

If waste is the issue, then a suggestive starting point for dealing with it may be found in the definition of waste itself. The Oxford Thesaurus equates "waste" with "expenditure" and "dissipation." How might a theoretical agenda for reconstructive design be derived from these notions? In classical thermodynamics, the dissipation of energy in a mechanical system is regarded as "waste." Ilya Prigogine changes this view by showing that dissipation becomes a source of order in open systems (Prigogine and Stengers 1984: 131–76). Waste as a source of order? This is indeed suggestive. Where in this seeming link between waste and order might we find a poetics – that is, rules of engagement – or even poetry? What is the relationship between expenditure in the operation of a gold mine and expenditure in a designed landscape system? The proposal for interventions in the Oceana Gold Heritage and Art Park uses the ancient formulation of the pleasure garden as a way to investigate these questions.

The work of the French philosopher Georges Bataille on the production of waste in human cultures provides a set of analytical tools for thinking about expenditure, as a critical practice, in relation to garden making. Bataille argues that, in an open system,

> the excess energy (wealth) can be used for the growth of the system. . . . [I]f the system can no longer grow, or if the excess cannot be completely absorbed in its growth, it must necessarily be lost without profit; it must be spent, willingly or not, gloriously or catastrophically.
>
> (Bataille 1988: 21)

What is of concern here is this glorious or catastrophic expenditure. It is in the very act, or practice, of expenditure that Bataille discovers a crucial element of becoming human. Human agents, enmeshed in social structures that he regards as necessarily legislative, regulative, rational, ordering, and controlling, require disturbance in order to tolerate that which is outside the cycles of possession, ownership, and profitable returns that characterize everyday life under capitalism. External to what Bataille calls the "restricted economy" of ordinary, daily existence there is an open realm of total loss, of unconditional expenditure, which disregards the self-preserving patterns of the restricted economy. This is the "general economy," a realm of unprofitable consumption, excess, annihilation, and birth. It is, in fact, the unknown, which the very laws and limits of the restricted economy aim to exclude. In short, this is Nature, or what Bataille refers to as the "sacred" (Bataille 1988: 19–23).

Now, as it happens, one of the points of the pleasure garden is to overcome the separation that exists between humans and nature. Paradoxically, however, the garden constructs a boundary or a frontier – an enclosing element – that immediately presupposes an excess that is excluded, that is not necessary for the garden to do its "work." That which is excluded from the garden is the mundane world of the restricted economy. From earliest times (since Persia and ancient Greece, for instance), pleasure gardens have been created to link humans to something beyond the everyday by separating them from the everyday (the Greek *paradeisos* refers to separation, or a terrain that is closed off). Gardens are, in other words, what Foucault calls "other spaces," that is, sites that are not subject to the order of ordinary existence, where use-value rules (Foucault 1997: 350–56). Gardens are declensions from this order. By their very nature, pleasure gardens are not utilitiarian.

Yet gardens have another, critical, dissipative function. Once marked off from the everyday by its boundary, the terrain within the garden wall is subject to a set of conditions that diverts it radically from the conditions outside. A series of operations – to do with such things as preparing the ground, drawing boundaries, and creating and maintaining horticultural difference – develops an internal dynamic as a function of technical practices, such as abstracting and grading the level plane, surveying and inscribing a boundary, spatializing the ordering system, and classifying and arranging plants. These systematic practices are put to the service, not of ordered production and maintaining existence, but of disorder. The garden frames and organizes the universe in such a way as to produce an imaginative realm where order is disrupted in a very specific way. In the pleasure garden, these organizational systems are brought into relationship with one another at the service of expenditure and dissipation, and that which is in excess of requirements is diverted into non-utilitarian aesthetic pleasure.

Conceptually, the Oceana Gold Heritage and Art Park is a dissipative structure. It has an ample prehistory of expenditure as a set of overlapping ecological structures, which produce waste as an automatic by-product of their natural functioning. Destabilized and subsequently transformed by mining, which is the "glorious expenditure" of capital and resources on the extraction of gold, the mine landscape became more complex as it became more disordered. Now, as extraction winds down, the possibility of connecting to new sources of order becomes evident.

The proposal

The first move toward reclamation is to avoid any intent to obscure the origins of the landscape in pre-human evolution, pastoral farming, and large-scale extractive operations. The second is to accede to a formulation of ecological succession that permits the recovery of former ecosystems, as well as the development of completely unpredictable new ecologies throughout the site. The third is to identify intensifications of the complex field of overlapping social, mineral, and organic phenomena that comprise the site. Such intensifications occur where many significant phenomena, which are both virtual and actual, coincide at specific locations within the landscape (such as a nineteenth-century miners' campsite, a native lizard population, and a lightning-struck Monterey pine). The Geographic Information System (GIS) program ArcMap can be used to identify these intensities on the basis of criteria, called "taxonomies," established for the purpose of reclassifying the landscape and, through this, finding specific locational qualities for actualization through design. The fourth move toward reclamation is to designate the intensities available for intervention by artists and landscape architects.

Nine land artworks have now been commissioned for the Heritage and Art Park, and three have already been constructed. Though these artworks are worthy of much discussion, I wish to explain here three of the intensities that have been identified for garden intervention.

Three gardens

Three gardens are designated for the town: a memorial garden, an archaeological garden, and a church garden. The criteria developed for the garden locations, or intensities, were as follows:

rocks + cultural memory + paradox = church garden
ecosystem + cultural memory + open space = memorial garden
entropy + commerce + heritage = archaeological garden

The church garden

At the western entrance to the town, a nineteenth-century Catholic church stands among mature *Pseudotsuga menziesii*, *Cupressus macrocarpa*, and *Pinus radiata* trees. It is located well back from the road, in its own space, and together with the large, dark trees forms a sentimental, and ambivalent, landscape element. Sheep graze the terrain around it, which slopes gently across the axis of the church; this land was once a stone field. Most of the rocks for the garden were collected when the church was built and were then incorporated into various structures throughout the town. Now the field and trees form a kind of sacred grove, into which the church is settled like an ancient temple, in every way a vivid reminder of the town's heritage.

Despite their associations with a specific time frame, however, historic landscapes are, like most other landscapes, emergent systems. Just as they emblematize a very human wish for fixity and persistence through time, they are nevertheless awash with novel matter-energy regimes. Even a well-preserved and well-conserved heritage landscape can never be a picture in a frame. Forever becoming, always transforming, such a landscape is subject to an array of turbulent inputs – mineral, biotic, social – that disturb its otherwise equilibrious state. It is therefore an adaptive landscape, and disturbance is crucial to its adaptivity and critical to its emergent character.

The proposal for a garden on this site introduces a sideways flow of matter-energy, subjecting its ambiguous historical charm to a destabilizing pressure and thereby transforming it into a new structure of increased complexity. It is a heritage landscape that evolves, but subtly (see Figure 3.3).

The windows of the church are redesigned as stained-glass artworks. The interior becomes a naturally lit gallery for art. Outside, enormous spalls recovered from the blasting of mines and pits are arranged in asymmetrical groups that drift across the meadow between the church and the road. These rocks are sufficiently buried so that, in most cases, they appear just to break the surface. Perennial flowers and herbs are planted in such a way that they flow among the rocks in waves of color and scent. A fractured-stone path winds from the rock outcrops to the church, and this mineral geometry forms a habitat structure for the area's native lizard population. The whole scenario is backlit from beneath the trees and behind the church. In open systems, there is a paradoxical closeness between structure and order, on the one hand, and between structure

Figure 3.3 Church garden.

and dissipation on the other: the landscape of the church is re-ordered by the introduction of mineral elements, themselves the products of the gold extraction's dissipative system.

The memorial garden

At the end of the main street opposite the church, at a bend in the eastern approach, there is an open space, anchored at the street side by three Lombardy poplars and bisected by a stream. Apart from the poplars and the stream, the site is "empty." A stone wall encloses it from the road, and there is a small gate. The space is slightly elevated and affords a prospect of adjacent fields, also inhabited by poplars and a stream.

Bataille's two systems – the general and the restricted economies – exist in a non-dualistic relationship that permits mutually effective crossovers between the two. These two economies, also identified as the everyday and the sacred, are linked through practices of excess and transgression, through what Bataille calls the "gift," a notion based on the concept of potlatch in pre-industrial societies. The gift can rupture closed systems by crossing their boundaries, interrogating the limits imposed on life by closed structures of exchange, legislation, hierarchy, and rationality. A society's memorial gardens are sacred terrain, sanctuaries that are separated from the surrounding territory by visible boundaries. A memorial garden has a protective character, a sense of asylum and of care and even of propitiation, certainly of atonement, respect, and celebration. In such a place – a space apart – we commemorate those who made it possible for us to be who we are and, in so doing, to mourn their loss. The process of commemoration connects us to their world. As a place where miners are remembered, the Macraes Flat Memorial Garden serves the purpose of the gift. The mourned are the men and women who worked and suffered for their work. In the early days of gold mining, the ore was taken by pickax and shovel and other rudimentary mechanical devices. Through their work they diverted a landscape system from its natural processes. More importantly, they sacrificed their sovereignty as individuals to the labor of gold mining. A landscape reclamation program should, in kind, divert thoughts to these people and a celebration of their labor.

Located at the northern entrance to the town, the memorial garden at Macraes Flat attempts, through specific features, to reestablish an intimacy with nature as atonement for that which is lost when human consciousness subjects itself to work, utility, and ultimately thinghood (see Figure 3.4). Individual gold miners, their families, and their communities are commemorated in

Figure 3.4 Memorial garden.

the design of the garden by means of their names, which are etched into the honed and polished surfaces of stone benches. These benches, made of local schist and polished to a marble shine, act like altars in a sacred grove, symbolizing sacrifice and loss and providing a focus for atonement. Lombardy poplars, already present throughout the region, strike emphatic, dark vertical forms against the western sky and dramatize the horizontal lines of the benches, which are arranged in staggered rows. A stream winds through the center of the garden, dividing a large space into two smaller ones. One half is "empty," the other "full."

The archaeological garden

There is another "empty" site within the town. It abuts the main street, between an old fire station on one side and the ex-Rabbit Board Headquarters on the other. This is the site of the former Old Kingdom Hotel and bakery, long since burned to the ground. It has become a garden too; however, its purpose is pure pleasure. The garden looks out over a wide landscape, in the foreground of which is a wetland that has been developed out of the previous agrarian landscape, once drained for productive purposes but now a recovering ecology. Beyond the wetland are huge stacks of rock and mine-tailing mounds, which create a sublime backdrop to the wetland. Vast land artworks are being created on these man-made topographies. The new pleasure garden, then, opens onto a complex cultural landscape. Within the garden, a spatial framework is constructed, in which a *locus amoenus*, a garden of delight, of transport, is realized. Its theme is material history.

Here, an archaeological feature is established and presented as a pleasure garden (see Figure 3.5). A square is laid out around the site, and the ground within the square is excavated to approximately two and a half feet below grade. A stone retaining wall separates the two levels thus formed and articulates the shape of the "dig." Any archaeological features that are revealed by the excavation, such as building foundations, remain within view as semi-revealed structures. The rest of the ground plane is leveled for safe walking, picnicking, and general relaxation – and to emphasize the inchoate archaeological remains. Beyond the perimeter wall, plants separate the site from its context and make it a special, enclosed space, except at the rear, where it opens onto a view of the wider landscape. Within the archaeological garden are planted fruit trees grown from the seeds of heritage species found in the Otago region: cherry, apricot, and almond. In one corner is a fountain.

A pleasure garden offers an emotional actualization of a time and place. By means of operations specific to its location and its temporal moment, the pleasure garden concentrates and focuses the whole world of nature, including human nature. It places the wide world itself within the experience of the visitor in such a way that there is a merging of what it is to be human with what it is not to be human. It encourages a concern for the world by means of an affective tone

Figure 3.5 Archeological garden.

drawn from the world and directed towards it, which is realized in the merging of subject and object. The garden thus transmits and sustains a human relation to a separate and extraordinary domain that has been designated as sacred. Such a garden is a gift from a remarkable tradition of making, from a tradition of terrain modification in the name of pure pleasure and through the expenditure that pleasure entails; it is a gift of connection to the "open realm of total loss" to which Bataille refers. This is a tradition that landscape architecture has too often forgotten.

Conclusion

As an ancient artifact of expenditure, the garden provides a direction for the design of landscapes that are themselves the production of waste. By focusing on the operations of garden making and the primordial reason for gardens – the connection to the open – landscape architecture can find a poetics in the relationship between expenditure in the operation of a gold mine and the glorious expenditure of the pleasure garden. In such a seemingly simple act as the making of a garden, landscape architecture can call into human lives the qualities of something beyond those lives. As Lynda Sexson has said, "[I]n contemporary culture so much of our world has been 'contaminated' with the mundane that we hardly recognise the quality of the sacred" (quoted in Berger 2006: 33).

Here it is important to recall the American tradition of nature writing, in which this quality has subtly and persistently been investigated. Thoreau, Burroughs, Muir, Austin, Leopold, Carson, and Abbey all place, in their own ways and according to their own preoccupations and dispositions, an emphasis on an imaginative understanding of nature and on a vision of something that is, let us say it, sacred. This vision, which Frank Stewart situates at the heart of nature writing, "is a sober and attentive rhapsody that we cannot help but be obligated by" (Stewart 1995: 232).

Bataille's work makes it possible to consider the Macraes Gold Operation site as a landscape of the gift. This idea requires a rethinking of what it means to reconnect to nature. Bataille's "open realm of total loss" is a nature that is valorized as so much more than an overlapping set of nested ecosystems. The critical question for landscape reclamation is, therefore, how do we reestablish the connection with nature that was lost through the reductive procedures of resource extraction?

The Oceana Gold Heritage and Art Park has provided an opportunity to investigate this query through a consideration of the pleasure garden as a kind of turbulence within the dissipative structure of an extractive landscape. Though necessarily small-scale, the examples here discussed point to suggestive design scenarios for large-scale landscape remodeling. In fact, the vast land artworks that are currently underway across the Oceana Gold site emphasize and extend the affective tone established in the modest township gardens. They contribute in another way and at a much grander scale to the poignant undecidability of mood that the application of "turbulence theory" has actualized among the waste-rock stacks and open pits of the Otago gold fields of South Island, New Zealand.

The small-scale gardens themselves locate and dramatize the ambivalent cultural values associated with resource extraction by deriving a design strategy, or poetics, from a theoretical consideration of its procedures. In this way, the complicated process of atonement is engaged with a critical emphasis on the fact that atonement is a never-ending process, not an objective.

Note

1 One is here conscious of the important work carried out on reclaimed sites by Mira Engler. See her *Designing America's Waste Landscapes* (2004).

Bibliography

Bataille, G. (1988) *The Accursed Share*, trans. R. Hurley, New York: Zone Books.

Berger, A. (2006) *Drosscape: Wasting Land in Urban America*, New York: Princeton Architectural Press.

Engler, M. (2004) *Designing America's Waste Landscapes*, Baltimore, MD: The Johns Hopkins University Press.

Foucault, M. (1997) 'Of other spaces,' in N. Leach (ed.) *Rethinking Architecture: A Reader in Cultural Theory*, New York: Routledge, 350–56.

Prigogine, I. and Stengers, I. (1984) *Order out of Chaos: Man's New Dialogue with Nature*, New York: Bantam Books.

Stewart, F. (1995) *A Natural History of Nature Writing*, Washington, D.C.: Island Press.

Chapter 4
Mines and design in their natural context

Dorion Sagan

>Destruction takes place so order might exist.
>Simple enough.
>Destruction takes place at the point of maximum awareness.
>*Orate sine intermissione*, St. Paul instructs.
>Pray uninterruptedly.
>The gods and their names have disappeared.
>Only the clouds remain.
>
>(Wright 1997: 62)

The philosophers Gilles Deleuze and Felix Guattari, in *What Is Philosophy?*, argue that, while intellectually separate, the great realms of art, science, and philosophy overlap one another (1996). The oldest form of art, according to Deleuze, is architecture, and, like science and philosophy, great art is marked not by rational and comprehensive closure, but rather by an openness and interaction with the larger universe ultimately constitutive of the artificial human divisions.

I concur with these comments and here attempt to apply a general understanding of thermodynamics to the context of mine reclamation. From the broadest possible "deep-ecological" perspective, the following pertinent things can be said:

1. Pollution is evolutionarily natural, a thermodynamic result of the growth of complex systems.
2. Living systems are open and, because they manipulate solids and work resources to build up their bodies and environments, may be considered "technological" from their inception.
3. This Ur-technology of Earth life is unconscious and biological, and completely natural.
4. Human-fostered technology, the productive part of which is the result of conscious planning, needs to integrate its idea of both technological production and pollution into a biological, evolutionary-ecological model that links conscious human projects to the unconscious "designs" of life in general.

Design in a natural context

How we make habitable, or at least educational and beautiful, the toxic husks of industry that are abandoned mines – into which approximately a dozen campers fall and die each year – is the task of reclamation. I argue that industry and technology, despite the tendency to see them as uniquely human, have deep precedents in nature. While we link form to art or aesthetics and function to biology and technology – considering their marriage of the two to be the result of human excellence in engineering or design – form and function in fact show up in absolute unity in completely undesigned, unthinking systems. These naturally complex systems, from which life most likely derived (Sagan and Schneider 2000, 2005), have functional mandates based on the laws of thermodynamics.

Thermodynamics provides a natural context for understanding design. Although we think of technology, design, and engineering as human and conscious, similar, exquisite systems in nature, including of course our own bodies, are unconscious. Think of the internal workings of the body beneath your skin that is the object of study of medical science. Although not perfect – otherwise there would be no medical science – the body, from capillaries to the optical and thinking systems, is a fantastically functional nanotechnological system that was not engineered by humans. Considering its many malfunctions, one may concede the imperfection of its design. (Indeed, its "peculiarities and oddities," as Darwin emphasized, are best explained as vestiges of once-functional systems, rendered useless as organisms changed to live in new environments.) The body is rather the result of natural growth and evolution, and these things are tied together by thermodynamics, the study of energy flow.

Living matter is an example of the general tendency of matter to organize, engage in cycles, become complex, and grow in areas of energy flow. This complexity and growth does not contradict the Second Law's mandate for increased disorder and equilibrium, but enhances it. Human technological civilization, to take a prime example, is a prodigious user of energy, displaying significant complexity even as its wastes can be measured in the atmosphere and global warming. Order and disorder are linked.

Nonequilibrium thermodynamics is the study of open systems. In thermodynamics, an open system is one that exchanges both energy and matter; a closed system (somewhat confusingly) is a system that exchanges energy (energy moves through the system, while matter does not); and an isolated system – the sort originally studied by thermodynamics – is closed to both matter and energy transfer.

Traditional thermodynamics began in the nineteenth century pragmatically as a concerted effort to hone efficiency in machines at a time when locomotives, as well as steam engines powering ships in the British navy, represented the height of technology. Thoughts in thermodynamics would help lead to nascent ideas in information theory and thus, in a roundabout way, would help lead to computers.

The kind of thermodynamics that studies open systems, nonequilibrium thermodynamics, developed later, not because such systems are less interesting or common – in fact, they are more interesting and more common – but because they are more difficult to study. Thus, this seemingly specialized subdiscipline really studies the more general situation of systems open to flux in energy and matter (Rosen 2005).

The Second Law versus complexity and design

The Second Law of thermodynamics, which says that entropy (a measure of atomic disorder, or "chaos," in the original, non-deterministic sense of the word) inevitably increases in isolated systems, is based on the simple observation that heat inevitably flows into the cool and not the other way around. The discovery of this principle upset the Newtonian clockwork cosmos, and suggested that the entire universe was going to wind down, run out of gas, and come to a cosmic standstill. Usable energy dissipates, making perpetual motion impossible. The discovery was made in the context of improving the efficacy of steam engines. One might argue that Murphy's Law – everything that can go wrong will go wrong – is a pop algorithm of the Second Law. Other versions of the Second Law include Matthew's Maxim, "If you think things are mixed up now, wait a while," and Roberts' Axiom, "Only errors exist." But if screwing up is so common in a cosmos that naturally tends towards randomness, then can natural functional complexity and design exist?

The answer is surprisingly simple. No architectonic consciousness is needed to actively oversee functional designs in nature. Rather, they "just happen." The reasons are described by nonequilibrium thermodynamics, and reflect the ability of complex open systems to produce entropy more effectively when they are more organized. Natural cycling systems that display complexity in three dimensions are better at accomplishing the Second Law's mandate to produce randomness than are mere random concatenations of matter. All energy systems run off gradients: temperature, pressure, or chemical differences in the environment. A steam engine, for example, runs off the thermal difference between a hot radiator and a cool boiler; a tornado comes into being in order to rectify a barometric pressure difference; and life taps into the electromagnetic difference between the high-energy sun and low-energy space. In each case, the energy-rich difference in the environment is broken down more effectively for the existence of the complex system. It is ironic: natural open systems, such as ourselves, which look like they are engineered, produce disorder more effectively, laying their environments to waste more massively (producing more heat and entropy), than non-complex systems.

Ecologist Eric D. Schneider, formerly of the National Oceanic and Atmospheric Association, a disenchanted district head of the EPA in Chesapeake Bay, provides a version of the Second Law that is more general than traditional versions, which incorporate the entropy measure and apply only to isolated systems. Schneider's version, which applies to open systems as well as to the sealed systems originally studied by thermodynamics, is that "nature abhors a gradient" (Schneider and Kay 1994a). As stated, a gradient is a difference across a distance, be it of temperature, pressure, or chemistry. Such differences, as in the vacuum that nature proverbially abhors, tend to be equalized. Life, far from being an exception to the Second Law, is one of its most fascinating manifestations (Schneider and Kay 1994b).

A Copernican Revolution

Science has taught us that, whether we like it or not, we are not so special. That four major upsets to the human ego can be identified as scientific progress has shown us that, far from being outside the universe and central to its operations, we are peripheral and ordinary. The first is the Copernican Revolution proper: we are not the center of the universe. The second is the Darwinian revolution: we are not the central organism, the be-all and end-all of life. The third is dual, reflecting both the discovery by chemists that life is made of ordinary chemicals, not any vitalistic stuff, and of astrophysicists that the elements found in life are not only not special but among the most common in the cosmos. The fourth such revolution, I would argue, is nonequilibrium thermodynamics: that, despite our capacity for thought and technology, the energetic processes in which we are involved are typical for a complex system growing in the area of gradients. Thus, just as the chemists showed that the stuff of life is not special, and the astrophysicists that organic compounds exist in space, we are now finding that the process of life is not unique. Life is first and foremost an energy-using system. We are thermodynamic systems. Our thought, behavior, and "design" are intimately related to the ways of energy. Our thermodynamic function is to find energy sources to power our own maintenance and growth. Such growth, especially in the absence of recycling, can lay the environment to waste, causing problems for the complex systems themselves. This is not unusual for growing organisms. The toxic waste we see in abandoned mines is one example of this natural, thermodynamically mandated process of resource extraction and waste. In evolution, major pollution problems have typically been solved by the appearance of new life forms that are able to use the wastes produced by growth. For example, free oxygen, originally a waste product of cyanobacterial growth, became tolerated, then actively used by other bacteria. The overcoming of oxygen's original status as a pollutant is part of the exciting history of life's cellular evolution (Margulis and Sagan 1997).

"Tyger! Tyger! Burning bright / In the forests of the night / What immortal hand or eye / Could frame thy fearful symmetry?" writes William Blake, who also wrote, "Energy is eternal delight," "The Road of Excess leads to the Palace of Wisdom," and "Energy is the only life." The answer to Blake's question, from a thermodynamic point of view, is energy and its thermodynamic, complexity-fomenting ways. Britain's industrial supremacy was founded on its extraction of coal and iron, useful in making machinery to extract further coal and iron in positive-feedback cycles. The interdisciplinary Soviet scientist Vladimir Vernadsky – who is on Russian postage stamps – has described life as a moving mineral. One of his biospheric principles is that over evolutionary time more and more elements become involved in biological circulation. Reclaiming poisoned mines is a fascinating human challenge, but detoxification – like its opposite, pollution – is not a singularly human challenge. The extraction and remediation of resources and energy has been going on since long before man's arrival.

Natural systems combining form and function

Long before life or landscape architecture, nature was in the business of forming energy systems whose designs were tied to their function. Consider Bénard cells, hexagonal flow structures in fluid (e.g. sperm-whale oil, silicone, sulfur-hexafluoride gas) that more effectively conduct heat than unorganized fluids. Henri Bénard had heard, in 1897, of polygonal flow structures arising in an unused photographic-developer tray. Fueled by a temperature gradient, the symmetrical hexagons arise out of the chaos. As the temperature gradient increases, a better way of dissipating heat occurs. It is crucial to note that as we go from conduction to convection, from random to organized, the system's heat flow – its entropy production – increases. Counterintuitively, the more complex, organized system is also the best at accomplishing the Second Law's task of disorder. The structure of the Bénard cells can be seen when the heated fluid is sprinkled with aluminum flakes, then photographed over a ten-second exposure: their form, the hexagons, comes from many individual trails of cycling fluid. The function of the cycling fluid is to dissipate the thermal difference. Bénard cells are a clear example of form tied to function in a natural complex system that may even be said to have an unconscious purpose – to reduce a gradient.

A more familiar example of natural form's tie to function is a storm system, the natural result of an atmospheric or barometric air-pressure gradient. Tornados and hurricanes swirl into being to reduce thermal and air-pressure gradients. That is their function – why they exist. Nature abhors a gradient and, where and when possible, will craft, without hands or brains, complex cyclical processes to get rid of them. This is the naturalistic basis of engineering and design: not mere mechanical grace imposed from the outside by a deistic intelligence, but a spontaneous coming together of systems to degrade gradients. The purpose of the tornado or hurricane is to reduce pressure gradients, and, in doing so, it produces waste as entropy in the form of heat. Aristotle wrote, over two thousand years ago, "It is absurd to suppose that purpose is not present because we do not observe the agent deliberating" (McKeon 2001: 251). It is arrogant to think that the universe makes things the way we do, by assembling them piece by piece. When we look at nature, we see structures come together as wholes, cycling matter as they funnel energy and dissipate gradients. Our mechanics are part of a larger, more unconscious form of engineering.

Chemical systems whose form is related to function include the famous Belousov-Zhabotinsky (B-Z) reaction, the final reaction of which is the oxidation of ceric acid by potassium bromate. These chemicals feed on a redox gradient, the same that powers your brain cells, which communicate as gradients are reduced in waves, as action potentials travel along axons, depolarizing electric charges across membranes. Building up these gradients requires food, blood sugar in the brain, which is oxidized. Although B-Z reactions are not alive, they, like surface life – including the metabolizing cells of the human body and brain – depend upon a chemical gradient. Life differs from nonliving complex systems in its ability to actively seek out energy sources to power its reproduction, which serves to continue its energy-transforming function in new systems after previous ones are exposed to damage, death, or decay by the same Second Law of thermodynamics that is its original raison d'être.

The challenge to landscape architecture

Life is a mineral transformer, an energized colloidal form of what Vernadsky calls animate water, which deploys its own stored energy in order to access further resources in a continuous thermodynamic process. Mountains and all-natural slag heaps are made from the carbonate skeletons of microbial organisms. The White Cliffs of Dover, sung about by Jimmy Cliff, are, for example, a stockpile of calcium-carbonate skeletons, a microbial mass grave. What begins as waste, after life returns to it, can become beautiful. One organism's waste can be another organism's food. Indeed, such alliances are at the heart of ecology and symbiosis. They are also embedded in our bodies. Calcium ions were poisonous for our two-billion-year-old free-living eukaryotic ancestors. The calcium had to be sent out of the cell. In some cases, as organisms became sexual and multicellular, the calcium was stockpiled to make shells or bones, that is, to make infrastructures and houses from natural thermodynamic waste. The next time you see somebody smile remember that it is a great feat of nonhuman engineering. The calcium white was once a microbial superfund site. Life has encountered pollution problems before. Evolutionarily, we have a lot of promise, but we are just new kids on the ecological block.

Autotrophy is a biology term that means "making your own food." The primary exponents of this lifestyle are plants and their photosynthetic forerunners, algae and cyanobacteria (a.k.a. "blue-green algae"). The chemical gradient formed by free oxygen at the Earth's surface is actually a photosynthetic by-product of the solar gradient. An estimated two billion years ago, bacteria that were already photosynthetic – but were using the hydrogen sulfide burped from volcanoes to get the hydrogen needed for their bodies – mutated to use the hydrogen in water. This move to water as a source for hydrogen in the making of organic hydrocarbon compounds resulted in the release of free oxygen into the atmosphere. Cyanobacteria thus oxidized the entire surface of the planet in a process that began two billion years ago and left a record of oxidized minerals on the Earth's surface. As the first to produce originally toxic (because very reactive) oxygen, they were among the first that had to adapt to the challenges of this new waste – the product of their own growth – in the environment. The analogy to human industry is germane: we, too, as rapid growers, have polluted the environment, especially the parts of it in which we reside. This requires that we either find ways of detoxifying those hazardous (if naturally so) places or reduce our own growth (biological and technological), or do both.

Interestingly, cyanobacteria themselves may be part of the answer for mines. Cyanobacteria – which can deacidify runoff water from mines – are guilty of a primordial pollution crisis. At Davis Mine in Rowe, Massachusetts, the runoff through the mine tailings is extremely acidic, at a pH of 2.5, but where there are gelatinous patches of unidentified bright-green algae, the pH goes to 6, which is less acid by a factor of ten to a hundred thousand times. If we compare the subsurface environments in which life seems to have evolved, along with the extreme environments in which microbes can survive, with the kinds of places we like to inhabit, it is clear that there is a big difference between what is hospitable to us and what is hospitable to life. In the dark, drenched in acid, life – including our cell ancestors – can thrive. But we are surface beings, adapted to sunlight and free oxygen, as are the cyanobacteria who turned this planet from the primeval equivalent of an abandoned mine into a realm of fresh air in the first place.

Bibliography

Deleuze, G. and Guattari, F. (1996) *What Is Philosophy?*, New York: Columbia University Press.
McKeon, R. (ed.) (2001) *The Basic Works of Aristotle*, Princeton, NJ: Princeton University Press.
Margulis, L. and Sagan, D. (1997) *Microcosmos: Four Billion Years of Microbial Evolution*, Berkeley, CA: University of California Press.
Rosen, R. (2005) *Life Itself: A Comprehensive Inquiry into the Nature, Origin, and Fabrication of Life*, New York: Columbia University Press.

Sagan, D. and Schneider, E.D. (2000) 'The pleasures of change,' in Daniel B. Botkin (ed.) *Forces of Change: A New View of Nature*, Washington, D.C.: National Geographic Society, 115–26.
—— (2005) *Into the Cool*, Chicago, IL: University of Chicago Press.
Schneider, E.D. and Kay, J.J. (1994a) 'Order from disorder: the thermodynamics of complexity in biology,' in M.P. Murphy and L.A. O'Neil (eds) *What Is Life: The Next Fifty Years*, Cambridge: Cambridge University Press.
—— (1994b) 'Life as a manifestation of the second law of thermodynamics,' *Mathematical and Computer Modelling* 19: 25–48.
Vernadsky, V.I. (1998) *The Biosphere: The Complete Annotated Edition*, M. McMenamin (ed.), D.B. Langmuir (tr.), New York: Springer-Verlag.
Wright, C. (1997) 'Sitting at dusk in the back yard after the Mondrian retrospective,' in *Black Zodiac*, New York: Farrar, Straus and Giroux.

Chapter 5
Ecological succession and its role in landscape reclamation

Eric D. Schneider

Presently there is in excess of 200 billion dollars being projected for or spent on ecological-reclamation projects around the world (Cunningham 2002: 1–6). There is the projected 20 billion for the restoration of California rivers and deltas, 12 billion for the restoration of a single watershed in China, and the 7.8 billion already spent (and the total is rising) for the restoration of the Florida Everglades. Also contributing to this sum is the cost of many thousands of small projects, from mine-water mitigation to stream restoration by small environmental groups like Trout Unlimited in the United States.

Most scientific journals and the ecological-reclamation industry point out that ecological rehabilitation and restoration projects want to produce or reproduce "natural" ecosystems. I ask: what are "natural" ecosystems? With this new giant industry, what are the measures of success? How do we know when we have produced a functioning ecosystem? In the following pages I will examine characteristics of natural ecosystems as they grow and develop through a process of maturation called "succession."

In a recent book, *Into the Cool*, Dorion Sagan and I made the case that much in nature is dominated by dynamic structures that "eat" high-energy material, extract from it the energy needed for subsistence and reproduction, and expend lower-grade energy into the surrounding environment (Schneider and Sagan 2005). This principle holds for cyclones, candles, cats, and cucumbers. Physicists call these systems "dissipative systems," because they dissipate energy in their process of existence. Here is where physics-thermodynamics meets biology and a new way of looking at life emerges. Thermodynamics shines a bright light on how ecosystems work. Jungles are not only beautiful repositories of numerous tightly interconnected species, but they are also the most effective living energy-gradient reducers known. These climax equatorial ecosystems cool the planet.

Taking high-quality energy from the sun, plants recycle the chemical elements required for life – hydrogen, carbon, oxygen, phosphorus, nitrogen, sulfur, and a few others – storing energy in biomolecules as they spin out the protein structures of sensitive living beings. While ecosystems, by definition, recycle most of their elements within themselves, the span of ecosystem recycling can be quite wide, and ecosystems can overlap. Life's nested cycling networks range in scale from cell metabolism to animals replacing cells in their tissues to the growing, living global-ecosystem that is the biosphere.

The sorry state of ecology

Ecology, as a science, has some problems. Mired in politics, it is associated with environmentalists, anti-globalists, nature lovers, and even eco-terrorists. Moreover, ecology's predictive power has been minimal. The state of predictive ecology has recently played itself out in the 10-million-acre Yellowstone ecosystem in the northern Rocky Mountains. In 1995, wolves were reintroduced to the ecosystem with little forewarning of the trophic cascade that would follow. It has been a little longer than a decade since the introduction, and many direct and indirect ecological effects have already been observed. One of the more surprising of these has been the 50 percent decrease in coyote populations in large parts of Yellowstone National Park. The wolves kill the coyotes, which are competitors in their ecological niche. Wolves generally eat only 50 percent of their kill, leaving the carcass for scavengers such as grizzly bears, ravens, and coyotes. Grizzlies have been known to give up their winter hibernation unexpectedly when food becomes available from wolf kills during the winter months. Ecologists knew that the wolves would impact the elk populations, since elk were expected to be the main food source for this apex predator. What was not expected was that the elk would move away from the exposed riparian areas (i.e. the green, vegetated areas beside streams and rivers) where they had nearly wiped out the aspen and cottonwood groves by overgrazing. In the Park's Lamar Valley, thriving young aspen and cottonwood trees now provide food and shelter for beavers, which have long been absent from the Lamar River. These animals make dams, which attract muskrats, otter, moose, and ducks taking advantage of the slow-moving water. This local increase in diversity is just one of the unpredicted top-down cascading effects of the addition of a single species to an ecosystem. This experiment has only just started, though, and the results are preliminary. The take-away lesson for those attempting the restoration of ecosystems is to expect the unexpected! A simple ecosystem of just twenty interacting species generates on the order of 2.4×10^{18} possible direct and indirect connections. So I say again: to a large extent ecology is not yet a predictive science and what we do know is the result of observation of nature and little from theoretical prowess. When one tries to re-create an ecosystem, one may get something different than what one expects.

Nonetheless, ecology is a genuine science. Global-diversity maps have been developed, showing diversity gaps as well as "hot-spot" regions of unusually high diversity. Population ecology studies species interactions, fisheries ecology accounts for fish stocks and their environments, ant ecologists study ant populations, and so on for other ecologists who concentrate on individual species, environments, and kinds of life. There is mountain ecology, deep-sea ecology, evolutionary ecology, human ecology, autecology (which studies how populations respond to environmental variables), the ecology of micro ecosystems, and microbial ecology. Technically, energy-based ecology is one of these subdivisions. It is within the "subdivision" of energy-based ecology – the thermodynamics of life – that the overall forces shaping ecosystem development may be understood. Indeed, the biosphere is an ecosystem, which means that the discussion of how energy shapes ecosystem development impacts ecology at its most inclusive, planetary level.

Alfred Lotka, Eugene Odum, Howard Odum, Ramón Margalef, Jeffrey Wicken, Robert Ulanowicz, and James Kay are among those who have shown how energy flow guides ecosystem development, how physical principles underlie evolutionary trends. Earlier biologists had glimmerings: French evolutionist Jean-Baptiste-Pierre-Antoine de Monet de Lamarck wrote of the "power of life," and English philosopher Herbert Spencer argued for a "universal law of evolution" toward increasing complexity, based on energy. Yet these theories were intuitive or philosophical rather than empirical.

Successions

Knowledge of the deep influence of energy on life began with simple observations. In the past 200 years of watching landscapes change, many patterns have been found. One of the first to describe floral-landscape changes over time, or succession, was Henry David Thoreau, at the 1860 Middlesex Agricultural Society annual cattle show in Concord, Massachusetts (Worster 1979: 70–71). A keen naturalist, Thoreau chronicled how abandoned fields became grasslands, then shrubs, and then pine and birch trees. These fast-growing, shade-intolerant trees then gave way to long-lasting hardwoods: forests of oak, hickory, and maple. After repeated observations, it appears that this successional sequence, which in New England occurs over 150-year periods, is predictable.

Since then, many variations on succession have been discovered. In each case, rapid growers settle an area. They are followed by new species. The recent arrivals need the fast-spreading pioneers that precede them. As waves of new species immigrate, diversity predictably increases. The ecosystem enlarges, and the rate of growth slows. If the ecosystem were merely the stochastic result of motley agglomerations of organisms fortuitously wed by common time and place, such concerted activities would not be expected. It is, moreover, difficult to argue that the regularities are genetically grounded. Ecosystems vary greatly in their genetic constitution, yet they display the same tendency to grow to a limit. If one looks at them thermodynamically, their common behavior makes sense. The ecosystem at maturity represents a system that has explored all possible "easy-money" dissipative routes and is now plodding along with the optimal amounts of energy captured in, and degraded by, its system.

Successions are everywhere. Not only is there the fallow-field-to-forest succession, but there are also the successions that take place in mud puddles, deep-sea vents, and the Amazon jungle. In the oceans, successions are distributed spatially, from the surface to the depths. The fast growers (plankton) live at the surface, while the slow-growing predators live at the depth. This is a vertical succession. The flora in a human mouth changes after each brushing, and species changes take place between brushings. A bacterial succession starts to take place in your mouth as the electric toothbrush knocks down the older species that have developed over the past twelve hours. It is like a chain saw in a fir forest, starting a succession that lasts 100 years.

The goal for those engaged in ecological restoration or reclamation is to start this successional process. Not only must one provide the species necessary for succession, but many niche constraints must also be fulfilled. For a stream or a pond, the water must not be toxic, or the succession will be arrested. The correct nutrients in appropriate ratios, the needed substrate (clay, sand, or rock), and the correct pH are just a few of the constraints that must be met for the ecosystem to thrive. The same goes for the restoration of a highly disturbed landscape. One cannot make a "bamboo forest" at 13,000 feet on unreclaimed toxic mine waste, so don't try. Again, the niche constraints – soil pH, toxicity, moisture, and so on – must be correct in order to get the successional process going. If one is starting from scratch with new soil, one must plant not only the stabilization crop, but also the seeds and plants of a future successional system.

A short course in ecology

There are volumes written on the science and history of ecology. Limited space dictates that I highlight the work of just two brilliant men: G. Evelyn Hutchinson and his student Eugene P. Odum. No one advanced the science of ecology as did these two men. Modern American ecology started at Yale University after World War II. Under the tutelage of the brilliant scientist G. Evelyn Hutchinson, modern ecology blossomed. His students and others attracted to his research program developed concepts such as trophic levels, the ecological niche, and, most important, the development of food webs and energy flow through ecosystems. Hutchinson's own work dealt mainly with species and populations ecology, and he made real advances in modeling simple ecosystems. He saw that there were both producers (autotrophs) and consumers (heterotrophs) carrying out the energy and material cycling of the biosphere. His students discovered food chains, whereby food energy is passed on when a predator eats its prey. They found and quantified the trophic pyramid, whereby many small organisms feed the few at the top of the chain. There are, for example, many more mice than cats. His students enumerated the species dynamics of islands and island chains. In short, Hutchinson set the science of ecology into motion.

After completing his undergraduate work at the University of Chicago, another man entered Hutchinson's ecological lair at Yale: Eugene Odum. After graduating from Yale with a PhD, Odum spent most of his life at the University of Georgia at Athens. He modernized ecology as a science, integrating it with geology, chemistry, and economics and management, and one of the most important syntheses of the science of ecology was Odum's 1969 paper "The Strategy of Ecosystem Development." Because the observations made in that paper and elsewhere play an important role in the present thermodynamic paradigm, it is worth taking a look at the process of succession as it might have been seen through his eyes.

Odum's classic text, *Fundamentals of Ecology*, first published in 1971, definitively linked energy flow to succession. In both his 1969 paper and his book-length text, some basic observations were made. With regard to ecological succession, he made three points: (1) "It is an orderly process of community development that is reasonably directional and therefore, predictable." (2) "It results from modification of the physical environment by the community." (3) "It culminates in a stabilized ecosystem in which maximum biomass (or high information content) and symbiotic function between organisms are maintained per unit of energy flow" (Odum 1971: 251).

Like Thoreau's work in the mid-1800s, Odum's synthesis of the changing properties of ecosystems during successions was phenomenological, drawn from centuries of observations. Data from both natural systems and from laboratory-sized ecosystems were used to develop the characteristics of Odum's database. He separated ecosystem characteristics according to the developmental stages expected during the development of ecosystems. He then grouped these characteristics into categories related to ecosystem energetics, community structure, organism life history, nutrient cycling, selection pressure, and overall ecosystem homeostasis. Odum's ecology fits comfortably into the present thermodynamic paradigm. I have attempted to synthesize Odum's basic ideas in Figure 5.1.

One of the most obvious features of ecosystem change during succession is the increase in biomass over time. The amount of biomass per square meter of early successional grasses is dwarfed by the biomass per square meter in a mature forest; the difference is one of grams versus tons per square meter. The biomass of an ecosystem increases and then levels off. The curve labeled B (biomass) in Figure 5.1 represents the ecosystem's finding and developing of new pathways for degrading available energy. The ecosystem grows. The more energy captured and flowing through a system, the greater will be the number and increase in entropy-production processes, such as transpiration, photosynthesis, and metabolic reactions, which all degrade incoming energy. In ecosystems, biomass, total system throughput, efficiency, and entropy production go up with succession and maturation.

Once the biomass curve levels off, some constraint has been applied to the growth of the system. This cessation of growth has multiple causes, ranging from genetic instructions to lack of light, water, or nutrients. If one thinks of the ecosystem as a thermodynamic dissipative system, at the climax, the system is in a quasi-steady state: it is organized to degrade all the available energy gradients as completely as possible, whether by autotrophic or heterotrophic means. An ecosystem is, in other words, a giant gradient reducer; the spread of organisms prior to the "climax" state creates the larger system, which is more intricate, complex, and organized – and better able to reduce gradients on a massive scale.

Using simple matrix algebra and measures of material and energy flow (who eats whom, primary productivity, defecation-nutrient data, reproduction, standing stocks of species, etc.), one can calculate detailed energy flows in an ecosystem. One can determine the total system throughput, the trophic levels, the proportion of flows involved in cycling, the number and length of cycles, and a wealth of other primary and secondary information on ecosystem states and processes. Although a computer can calculate these measures in a matter of minutes, the collection and enumeration of this data entail years of careful work. Not only do field biologists have to measure standard ecological data such as primary production and community and organism respiration, but they also have to examine the stomach contents or feces of each

Eric D. Schneider

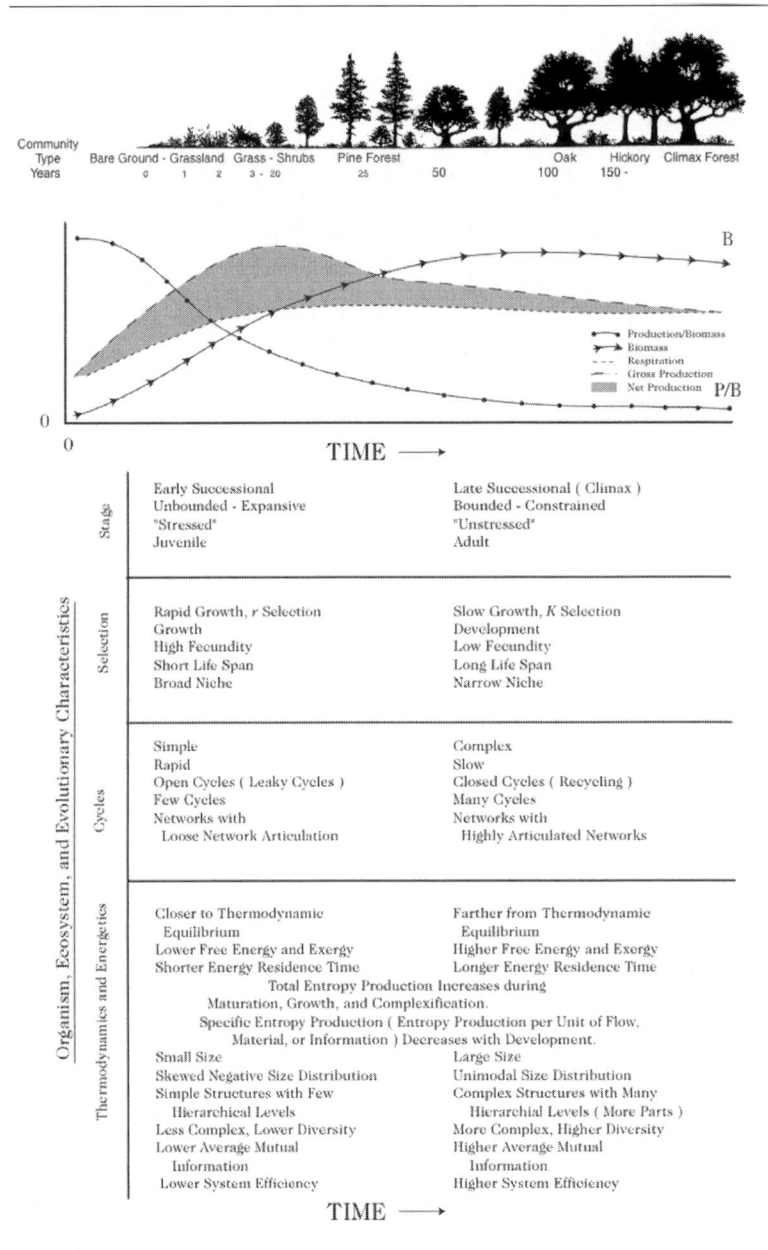

Figure 5.1 The diagram maps changes in ecosystem characteristics with development. Across the top of the figure is the flora associated with the archetypal New England succession, from fallow field to oak-hickory forest. In the western United States, the sequence would look similar, with sagebrush instead of blackberries and bunchgrasses instead of crabgrass. The final stages of the western forest would be fir, spruce, or white pine instead of oak. The same successional processes are seen in grasslands, where there are no forests of oak and fir. Grasslands go through their own succession, as annual grasses are replaced by perennials. The ecosystem is doing its "best" to prosper under its constraints. Many of the steppes and grasslands of the world are water-limited, so the grasslands have evolved to develop a stable community that degrades incoming energy as completely as possible.

Source: Schneider, E. and Sagan, D. (2005) *Into the Cool*, Chicago, IL: University of Chicago Press, Figure 13.2, p.199.

organism to find out what its diet consisted of. If the organism of interest is a grizzly bear, she could have eaten ants, white-bark-pine seeds, cutthroat trout, moths, scavenged elk, at least ten species of berries, or even an automobile oil filter.

A most important curve in Figure 5.1 is the production/biomass relationship over successional time. It represents the primary production (P) of a system divided by its biomass (B), per unit of area. This P/B ratio shows how much primary production is needed to support a unit of biomass and is an important measure of ecosystem health, successional progress, and system efficiency. Odum and Ramón Margalef have emphasized that this measure offers a sort of metabolic temperature for an ecosystem. P/B ratios decrease during ecosystem succession. Data from organisms and ecosystems show that, as these biosystems mature, their metabolic rate lowers; so does specific-entropy production. The lowering of the metabolic rate accompanies a leveling off in growth – growth that culminates in increased efficiency at maturity. The biomass supported per unit of energy flow decreases during succession. The system stabilizes. Like adult organisms, it can maintain itself with equal or less incoming energy or materials. P/B ratios are simple measures that field restoration ecologists can easily measure to give a quick assay of the progress of a reclamation site.

Odum describes food chains in early successional systems as simple and linear; more mature food chains are more complex, with more cycles, including the recycling of nutrients from detrital material produced by the ecosystem. The efficiency of transfer of energy low in the food chain from one species to the next is small (1–2 percent); however, higher in the food chain, say from rabbit to fox, this efficiency of transfer is greater (up to ~25 percent). It is hard to build a trophic pyramid of more than five or six levels, because of the inefficiencies in energy transfers at each trophic level. The second law of thermodynamics extracts its entropy tax at each of these transfers which results in a small number of trophic levels in ecosystems. Material can, however, be passed around in the heterotrophic detrital food chain many times. Late successional systems have much more complex food webs and many more pathways for the recycling of energy and materials (Schneider and Kay 1994: 38). The ecosystem may not have the distinct skin or bark enclosure of an organism, but it is an integrated system nonetheless. And it eats. Only about 1 percent of the energy coming to a plant is turned into living matter; the other 99 percent is either reflected back into space or turned into low-grade heat entropy. But that 1 percent is put through the paces. Mega-heterotrophs, such as cows, eat grass, and bacteria in their digestive tracts break down grasses until as much of the solar gradient is reduced as possible. The energy gets a double whammy: both autotrophic degradation and then heterotrophic processes – first plants, algae, and photosynthetic bacteria, then animals and their ilk – wring the last of the low-entropy products from the incoming solar rays. Solar energy is degraded at many places in the ecosystem.

Early-stage, r, ecosystems often lose nutrient resources, while mature ecosystems retain much of their resources within the system. Ecologists call material from outside the system "exogenous"; material that initiates within the ecosystem is called "endogenous." A rain forest recycles most of its material and energy, providing an endogenous source of water and nutrients for the system. Autotrophic leaves, containing cellulose, are attacked by herbivores; and by the time the heterotrophic food chain is finished with the leaves, litter, bark, and mulch, many of the gradients (chemical bonds) have been squeezed out of the system, and simplified raw nutrients are available for recycling. Such processes require well-developed root and heterotrophic organic environments where these transitions can take place. In the rain forest, even moisture is recycled. Early-morning and noontime temperatures, clear skies, and sunlight lead to afternoon showers – a rapid-response cycling system. Early successional systems do not have the root systems, leaf biomass, and organic material to make such recycling possible. A key attribute of a mature ecosystem is that it does not leak a lot of its nutrients, material, and water from the system as it recycles them.

Cycles are a ubiquitous ecosystem phenomenon. Cycling of material and energy within the ecosystem changes during succession. In early succession, the cycles are short, open, and fast. In mature ecosystems, the material and energy cycles are just the opposite: with long, complex cycles that are closed in upon themselves. It was shown earlier that more efficient ecosystems with low P/B ratios have the longest residence time of material and the most material cycling. A simple unit analysis of the P/B ratio shows that it is equal to the residence time ($1/T$), or the cycling time of material in the system. The early r species in a succession have a short lifetime; for example, nitrogen in a bacterial cycle has a residence time of hours. The residence time of nitrogen in a mature forest's ecosystem, on the other hand, is measured in hundreds of years. In these mature systems, there are not only more cycles, but also more material flowing through the cycles. In terrestrial reclamation sites, one should collect the runoff and a handful of easy nutrients to measure. If the system is maturing, both runoff and nutrient loss should be decreasing with time.

During succession, niches narrow. A "broad" niche refers to the great range of environments in which organisms can live and is typical of early-succession systems. A "narrow" niche evolves when competition demands that organisms develop traits and specialties to survive in a more complex world. Specialization ensues, and sometimes niches will overlap. The hermit crab, for example, uses a snail's abandoned shell, having developed a niche that is dependent on the biological hardware of a shell left behind by another organism.

Climax ecosystems

One of the most misunderstood concepts is the process of a climax ecosystem in a successional sequence. One problem with the climax concept traces to its early investigators, who saw the climax as the final steady state of a system. The error of this concept can be seen in very mature ecosystems. Large trees are vulnerable to wind throw, for instance, and overcrowded forests are vulnerable to fire and infestation by herbivore pests. Wooded areas with lots of biomass on the ground are forest fires waiting to happen. With each of these destructive processes, parts of the mature forest are driven back to organizational ground zero and start to rebuild their successional processes again. Most often, events like forest fires do not scorch immense areas at one time. The famous fires of 1988 in Yellowstone National Park skipped and hopped over the landscape, with only one-third of the 2.2-million-acre park burned.

A climax ecosystem should be looked at as a mosaic of different stages of ecosystem succession. Each time a major setback happens – a big forest fire, for example – the ecological structure and process are set back to the beginning, like a player sent to Go on a Monopoly board (the initiation point for the game). The penalty is not being thrown out of the game, but being set back to zero to start the progression over again. If, therefore, one were to look at the biomass curves over a long time, one would see a sawtooth pattern develop, as biomass builds up through thermodynamic growth, the material is destroyed, and slowly ascends to rebuild the more mature climax condition. Because of their patchiness, especially in disturbed areas, ecosystems tend to never die except when replaced by blacktop parking lot. But be assured: bacterial successions are still going on in the soil beneath. Ecosystems are not organisms, although, like organisms, they are complex thermodynamic systems. Ecosystems represent biologically stable patterns across many scales of a space-time hierarchy.

Regress under stress

Ecosystems regress under stress. Marine ecologist Kenneth Sherman and his colleagues cataloged species in ecosystems under pressure from commercial fishing (Sherman et al. 1981). The removal of larger, long-lived commercial fish associated with climax ecosystems resulted in a population boom of less valuable sand eels. Researchers linked the smaller, more quickly breeding sand eels to depleted catches of commercially important herring and mackerel. It is an example of succession being pushed backward. Abundant stocks of herring and mackerel typify the climax marine ecosystems beloved by fishermen. Removing these fish plunged the ecosystems to earlier stages of development marked by greater numbers of a faster-growing species. Such ecosystem reversal appears to be universal. Depriving ecosystems of sufficient energy or upsetting their interconnected integrity decimates their degrading capacities, physiologically forcing them back into states they had grown out of. When the energy available for the formation of complex systems is taken away, these systems revert to a more primitive level of function.

The clear-cutting of oak and fir is the terrestrial equivalent of draining mature fisheries stocks from the ocean. Since ecosystems are nested networks, stressing them does not usually kill them, but rather sends them back to an earlier stage of complexity, when they are colonized by more fecund species, the *r* pioneers of the successional process. Stress sends the gradient-reducing system back to earlier modes, which are able to make do with less energy.

Another example of an ecosystem's regression to earlier successional stages can be seen in an experiment using lethal levels of radioactivity in a mature ecosystem. George Woodwell and his colleagues at the Brookhaven National Laboratory bombarded a mature oak-pine forest with high levels of gamma radiation for ten years. Woodwell's group measured and photographed the effects of the bombardment. "The radiation," they summarized, "first destroys the pine trees and then other trees, leaving tree sprouts shrubs and ground cover. Longer exposure kills shrubs and finally the sedge, grasses and herbs of the ground cover" (Woodwell 1970: 20). This was Thoreau and Odum's ecological succession driven backward in time. Whereas succession captures and degrades as much energy as possible, radiation can reverse the process. In this case, the irradiated oak-pine forest retreated to an earlier stage in its development as an ecosystem. In the end, fast-growing *r* species and very hardy lichens dominated it.

All individual organisms are bounded by structures of their own making. A tree is enclosed by bark, mammals are encapsulated by hairy skin, and even gram-negative bacterial cells have walls enclosed in membranes. Ecosystems have boundaries. A mature ecosystem leaks very little of its nutrients and water.

At the Hubbard Brook Experimental Forest in New Hampshire, an experiment to measure how well stressed ecosystems maintain their materials began in 1965 and continues today. Yale University professors Gene Likens, F. Herbert Bormann, and their colleagues have been studying sites maintained by the U.S. National Forest Service. Likens's lifelong research program focuses on the biogeochemistry of forest ecosystems. His long-term studies at the Hubbard Brook Experimental Forest, which he co-founded with Bormann, have shed light on critical links between ecosystem functions and land-use practices. A watershed in the Hubbard Brook Forest was sprayed with herbicides after the woods were clear-cut in the fall and winter of 1965. For several years following this ill treatment, researchers monitored the water and nutrient flow through the drainage basin. They compared the results with those from similar drainage basins, in the surrounding woods, that had not been cut or sprayed (Likens *et al*. 1970).

The results were dramatic. The stream runoff – the watery leak – in the deforested system increased by 39 percent the first year and by 28 percent the second year. Blasted by pesticides back to a very early successional stage, the integrity of the ecosystem drastically declined, leaking its most valuable resource, water. Other valuable materials – phosphate and nitrate – were also lost at a much greater rate than in the undamaged basins. Nitrate loss increased forty-one-fold, which meant that nitrogen in the herbicide-treated area was far less available to organisms for making their proteins and nucleic acids.

Two years later, nitrate was still escaping at fifty-six times its rate of release from the undisturbed ecosystems. While the mature systems retained and recycled nutrients and water, the pesticide-damaged ecosystem was losing its grip. The floodgates relative to the healthy drainage basins opened, and runoff rates skyrocketed: 417 percent for Ca^{++} (calcium), 408 percent for Mg^{++} (magnesium), 1,554 percent for K^+ (potassium), and 177 percent for Na^{++} (sodium). The pH of the basin's stream also dropped: it became more acidic. With no trees to shade its banks, the temperatures of the basin's stream rose, and the water clouded as sediments washed downstream. Thus, the "stressed" ecosystem – the cut and sprayed watershed – leaked nutrients, water, and sediments. The more mature, uncut watersheds recycled these materials. The increased cycling of material and energy is a hallmark of a mature dissipative system.

Pesticide, radiation, and oil pollution impair ecosystems and cause them to malfunction. They no longer capture as much energy or build the intricate structures that they once did. They no longer expand along natural successional trajectories toward maturity.

The tale of two creeks

Three Howard Odum graduate students studied two adjacent tidal-marsh creeks on the southern coast of Florida in the 1970s (Homer et al. 1976). One tidal marsh had a 2,400-megawatt nuclear power plant on it; the other creek maintained its natural character. The energy from the power plant was tapped by turbines feeding on the temperature gradient between the water heated by nuclear reactions and the cool creeks outside. The Crystal River nuclear plant raised the temperature of the adjacent creek by 6°C. This increase in water temperature "stressed" the ecosystem in the creek. Although the plant was nuclear, the stress was heat-related; and it was not temperature per se, but the move away from ambient conditions – from the ecosystem's comfort zone – that damaged the system.

These hardworking graduate students analyzed who ate whom in the creeks and how much and when they ate. They measured the energy captured by the algae and large plants and traced energetic transformations into organic compounds. The stressed ecosystem sustained a 34.7 percent drop in its biomass. It became "sick," in a way, and "lost weight." So, too, the total system throughput (TST) went down almost 21 percent in the stressed system, representing a loss of overall gradient reduction capability. The stressed ecosystem regressed to an earlier, less functional state.

Even worse, the complex food web was severely compromised: the number of cycles was cut in half; the stressed ecosystem lost much of its recycling power. The hot water from the power station devastated the stressed ecosystem's ability to retain the material it had captured. It leaked nutrients and energy like crazy. It wasted away. From this example and those above, it can clearly be seen that stress sends ecosystems back to earlier stages of development. They function more simply, with less diversity. They shrink.

Conclusion

Ecology is a data-rich science. Hundreds of millions of dollars are spent each year to collect and analyze environmental data on ecological systems. Sadly, much of this data is incomplete and rarely synthesized. Often data are taken for a particular reason, without an experimental design that enables answers to larger questions. We have highlighted a few excellent data sets, well-thought-out experiments and sophisticated analysis, which have produced common threads that make up the fabric of ecology. The data sets detailed above show that early successional and "stressed" ecosystems have lower energy flow, lower efficiencies, fewer cycles, less material in those cycles, lower interconnectivity, and greater loss of nutrients and water than more mature, or "natural," ecosystems. Successions are an unfolding energetic process that builds structure and processes to capture available gradients and degrade that captured energy as efficiently and completely as possible. The job of a reclamation ecologist is to set these succession trajectories in motion. Using ecological principles, one may guide the reclamation of disturbed systems toward a planned outcome. Nearby ecosystems ought to be one's guide. Try to reproduce the ecosystemic conditions (in terms of moisture, soil type, and species composition) of adjacent systems, and then let the disturbed system find its own new stable states.

I appreciate the ideas and words of Dorion Sagan and the editing of Michael Cloud Devine.

Bibliography

Cunningham, S. (2002) *The Restoration Economy*, San Francisco, CA: Berrett-Koehler Publishers.

Homer, M., Kemp, W., and McKellar, H. (1976) 'Trophic analysis of an estuarine ecosystem: salt marsh–tidal creek system near Crystal River, Florida,' unpublished manuscript, University of Florida, Gainesville.

Likens, G., Bormann, F., Johnson, N., Fisher, D., and Pierce, R. (1970) 'Effects of forest cutting and herbicide treatment on nutrient budgets in the Hubbard Brook watershed-ecosystem,' *Ecological Monographs* 40: 23–47.

Odum, E. (1969) 'The strategy of ecosystem development,' *Science* 164: 262–70.

—— (1971) *Fundamentals of Ecology*, Philadelphia, PA: W.B. Sanders Company.

Schneider, E. and Kay, J. (1994) 'Life as a manifestation of the second law of thermodynamics,' *Mathematical and Computer Modeling* 19: 25–48.

Schneider, E. and Sagan, D. (2005) *Into the Cool*, Chicago, IL: University of Chicago Press.

Sherman, K., Jones, C., Sullivan, W., Berrien, P., and Ejsymont, L. (1981) 'Congruent shifts in sand eel abundance in western and eastern North Atlantic ecosystems,' *Nature* 291: 486–89.

Woodwell, G. (1970) 'The energy cycle of the biosphere,' in Board of Editors at Scientific American (eds) *The Biosphere*, San Francisco, CA: W.H. Freeman and Co.

Worster, D. (1979) *Nature's Economy: The Roots of Ecology*, New York: Doubleday.

Chapter 6
Interrogating a landscape design agenda in the scientifically based mining world

Belinda Arbogast

Introduction

Mineral-resource development in the United States continues to be essential in maintaining current standards of living. Our cities are built of stone, clay, steel, and glass. We cannot build without removing minerals from someplace else, but consumptive use of minerals involves difficult choices and a struggle between land use, land degradation, and natural-resource conservation. Assessments of potential mining impacts should be at the heart of mineral development and production. Examining past mining examples can lead decision makers and the public into understanding their present landscape and encourage innovation in future reclamation projects.

Wise regional development takes into account the mutual importance of science and technology, the social world (our human experience – past and present), and aesthetics. There are increased demands for environmental benefits (such as sustainability and biodiversity) and socio-economic benefits (including jobs, drinking water, housing, recreation). Land planners, industry, government agencies, and communities demand that more attention be paid to best management practices and landscape reclamation.

The landscape architecture profession is a source of creative thinking and planning that is relatively untapped within the mining industry in America (Dempsey *et al.* 1979: 36). This chapter describes two design issues in mining and reclamation: 1) the lack of dialogue between designer and scientist, and 2) the integration of landscape architecture within an earth-science organization. One of the main challenges to developing creative science-based designs for mining reclamation is the lack of interaction between scientists and engineers, who have traditionally dominated the mining field, and design professionals, who may feel that incorporating aspects of science and statistics interferes with the creative process. What is needed is the willingness to experiment and see what the world of design and planning has to offer the world of earth science and vice versa.

The landscape architect is a creative problem solver, yet there are no simple answers to the problems facing society's need for minerals, resource development, and mined-land reclamation. Land-use decisions and site design are based, at least in part, on science (or should be). The ability of government representatives, industry, scientists, designers, and the public to understand and incorporate science into the decision-making process is influenced by personal values, beliefs, and a knowledge of science. Although most Americans have some interest in science and technology, science literacy in the United States is low (National Science Board 2000).[1] In

Figure 6.1 The drawing is by W.A. Rogers, titled "A street in a mining village in Pennsylvania," 1888, courtesy of the Library of Congress, Washington, D.C.

addition, lawmakers, judges, and other administrators involved in making decisions about resource development and reclamation often lack scientific literacy (Faigman 1999: xi). This is an information gap that designers and scientists need to fill.

Education, technology, and the acquisition of spatial literacy are great equalizers for landscape architects and scientists. The use of geographic information systems (GIS) and verifiable data in site design, visual analysis, and resource management has increased. We are in a time and place where we can click on a map, view census data, retrieve watershed information, and create digital elevation models (DEM) from a personal computer. Acceptance of the landscape architect as a peer by scientists on an interdisciplinary team, however, can be a challenge.

If landscape architects want to be influential members of multidisciplinary teams, along with scientists and engineers, they must address the misconception of landscape architects as solely artists or horticulturists. Some governmental agencies, such as the U.S. Forest Service, National Park Service, and Bureau of Land Management, have landscape architects on staff: approximately two hundred, four hundred, and six, respectively (Cownover 2004). Scientists may recognize the value of visibility studies, but many remain unclear as to what other expertise a landscape architect can bring to a project.

The five individuals interviewed for this chapter include a geologist, a geochemist, a hydrologist, a biologist, and a botanist with the U.S. Geological Survey. Their jobs include assessing mineral resources, detailed geologic mapping, research on acid-mine drainage, and environmental problems related to active- and abandoned-mine lands. They raised three common questions regarding the profession of landscape architects:

1   What *do* landscape architects do?
2   What skills can they bring to the table?
3   Do they have the necessary background to be involved in environmental work?

These federal employees believed that the landscape architect lacks knowledge of local climate, drainage issues, topography, plant nutrition and application in problem soils, wetlands and other habitats, budgetary concerns, and pollution. Each of these issues is of concern in mining and reclamation. The botanist noted that geochemical knowledge is necessary for understanding the impact of soil conditions on revegetation at a reclamation site. She assumed that botany and environmental biology ought to be required studies in landscape architecture. All of the federal employees agreed on a need for the landscape architect to have a basic understanding of climatology, hydrology, geomorphology, soil science, plant ecology, native species, economics, and surficial geology in order to participate fully in reclamation studies. They believe that the landscape architect must understand science and think critically in order to create designs that satisfy basic environmental processes.

At the same time, the federal employees recognize the importance of cultural issues that scientists may discount, such as aesthetics, sense of history, or land use and zoning. The geologist noted that a simple pedestrian bridge, let alone the siting of an open-surface mine, can cause community uproar. In Evergreen, Colorado, a two-hundred-foot bridge, which provides a trail connection between a dam and an underpass, was perceived by the community as an eyesore, as contrary to the community's character, and as a danger to use when ice and snow fell. The design was also criticized for not following the site topography. The community had not been consulted on the design of the $600,000 bridge project. In order to create successful mining projects, landscape architects must ensure fundamental community involvement from the beginning. It is a mantra too often ignored.

Figure 6.2 Panoramic view of a gravel pit and surrounding landscape on the Wind River Reservation, Wyoming. Photo by B. Arbogast.

Figure 6.3 A landscape architect is taking field notes for a visual assessment regarding the proposed development of a sand and gravel pit. Photo by B. Arbogast.

Interrogating a landscape design agenda

The geochemist is a Native American and noted the significance of history, culture, and beliefs in resource development. She noted the importance of building relationships and trust. Respecting traditional knowledge, acknowledging the distrust of the government by some groups, attempting sustainable viewpoints, and recognizing diversity among cultures will aid in researcher acceptance by the community. Public values and attitudes in the communities and regions adjacent to industrial mineral resources must be explored in order to assess fully the impact of mining. Psychology and sociology can explain public opinion and give insight into different time periods and places. They are areas that earth scientists tend to avoid yet which landscape architects explore.

Landscape architects have the methods and tools to create a dialogue between science, mining, and society. Designers bring creativity, a different perspective, and unique skills to a team of scientists. Daring to take on expert knowledge and share ideas can be daunting in a room full of strong opinion and logic. Designer terminology, such as form, line, and genius loci, needs to be well explained, and jargon should be avoided. Imagination must have its feet firmly planted in the world of objective and verifiable fact. The landscape architect must be willing to question and aspire to objectivity.

In addition to the traditional design issues of public preference, sense of place, and aesthetics, the environmental impacts of mining need to be addressed. Mine sites are a perfect place for design and science to overlap. If the profession is to be involved in mitigating the impacts of a proposed quarry, the landscape architect must make recommendations that go beyond reducing the visual intrusion of a mine. Such recommendations require being familiar with topsoil (salvage, maintenance, and redistribution), slope stabilization, grading, soil analysis, revegetation (e.g. density, species-richness), fugitive dust and noise, surface and groundwater protection, wildlife habitat, waste disposal (including tailings), and drainage control (surface runoff, diversion structures). Scientific information, with proper visual representation and explanation, can be used in meetings and before the public to justify design decisions. Resolving disagreements and promoting interagency cooperation requires excellent communication skills, from written and verbal, to multimedia and audiovisual.

If the landscape architect is not familiar with basic earth-science facts, concepts, and vocabulary, how can he or she determine the extent and characteristics of surface disturbance by a mine site, or recommend design alternatives for post-mining use? Restoring ecosystems and biologic processes involves more than naming a plant palette. As the Canadian Society of Landscape Architecture states, "Landscape Architects must understand the roles of the various allied professions and develop skills to direct work and participate on a variety of teams to meet those needs and challenges" (1996: 1).[2] Critical thinking, time, and experience will mature the scientific design mind.

Figure 6.4 Reclamation work underway at the Howe Pit, outside Denver, Colorado. Photo by B. Arbogast.

Creativity and social issues in an information age

In order to understand and design mining reclamation in a complex multicultural world, designers need to be able to apply an integrated approach and to visualize landscape change over time and on a variety of scales. To this end, they need to be able to utilize a variety of data types, from cultural features (derived from ground-based photographs, remote-sensed images, and historic maps) to physical features (derived from topographic maps, geologic maps, and other GIS layers). To be aware of and able to collect such environmental information requires ongoing education.

In reclamation work, there is an opportunity to educate in the process of restoring or creating a new landscape. Even philosophy and art criticism come into play when evaluating mined-land reclamation. When science reaches beyond its frontiers, it merges with philosophy; likewise, art can be dematerialized, boiled down to pure ideas (Mayor 2001: 5). The landscape architect has an ability to interpret information, be a team player, define problems, evaluate alternatives, and facilitate stakeholders. Just as scientists do, the landscape architect can examine data, determine the facts, make an analysis, and publish a report.

Traditionally, national geological survey organizations (including the U.S. Geological Survey) developed information databases for use in minerals, fuel, and water-resource studies (Findlay 1997: 21). Geoscience information today goes beyond mapping and mineral exploration. There are rich databases, newly acquired satellite images, and a wealth of reports available to the designer, accessible through the Internet. The U.S. Census Bureau, U.S. Environmental Protection Agency, U.S. Geological Survey, U.S. Forest Service, Bureau of Land Management, and Office of Surface Mining have websites that provide educational resources, mineral facts, digital maps, photographs, and publications that can assist any research project dealing with the issues surrounding mining and reclamation.

The U.S. Geological Survey (USGS) is the nation's primary provider of earth-science information on mineral resources. How does a landscape architect fit into that organization? The USGS is exemplary of the increased social direction in national geological survey organizations (Findlay 1997: 1). Its mission is to provide the nation with reliable, impartial, and timely scientific information for describing and understanding the Earth. This information is also intended to "minimize loss of life and property from natural disasters; manage energy, mineral, water, and biological resources; enhance and protect the quality of life; and contribute to wise economic development and a sustainable future" (U.S. Geological Survey 2003).[3] In other words, the USGS is committed to investigating the processes that affect the surface of the Earth and society-driven issues related to quality of life.

The exploration, development, production, and reclamation of a mine raise environmental, economic, and aesthetic issues. The remainder of this chapter describes one USGS employee's efforts to investigate the environmental and visual impact of surface mining and reclamation as part of a scientific team.

A landscape architect in the U.S. Geological Survey – a case history

Approximately ten years ago, the USGS Chief Scientist for the Central Region Minerals Team, Geologic Division, gave a chemist the opportunity to merge the principles of landscape architecture with physical-science information. The task would take a multidisciplinary approach, identifying the environmental and aesthetic factors that are required to assess the effectiveness of existing and planned reclamation. It was a plan to step outside the box and approach mining from a different perspective. Because the scientist had past experience in chemistry and quality assurance, it was believed that there would be greater acceptance of the chemist's work in mining-related landscape architecture by fellow scientists. The following are examples of the kinds of products that can result from such integrated studies.

One of the first publications, *The Human Factor in Mining Reclamation*, addressed the prejudice that all disturbed landscapes are wastelands. Through an evaluation of resource-

Interrogating a landscape design agenda

management history, regulatory considerations, and varying definitions of landscape, nine design approaches to mine reclamation were documented: natural, camouflage, restoration, rehabilitation, mitigation, renewable resource, education, art, and integration (Arbogast *et al.* 2000: 13). The report is useful for specialists in the mining industry, as well as landscape architects and land planners, and it especially benefits smaller companies and agencies that may not have the resources to research innovative methods of mining reclamation. Such groups as county environmental services departments and the Colorado Division of Minerals and Geology also use it as educational material.[4]

Interest in 'A Comparison of Landscape Form and Change from Anthropogenic and Natural Earth Movement – Understanding the Public Perception', a poster at an industrial minerals forum, resulted in the presentation of another paper (Arbogast 2001 and 2002). The delegates consisted mainly of geologists and engineers from various geological surveys, private mining companies, and industrial-mineral consultants. They were intrigued by the subject matter, understood its relevance for the industry, and enjoyed the visual quality of the presentation. Each publication, each research project, is a step toward gaining respect and repute from fellow scientists for the profession of landscape architecture.

The geosocietal factors of industrial-mineral development and reclamation are explored in more detail with the multidisciplinary work 'Evolution of the Landscape along the Clear Creek Corridor, Colorado – Urbanization, Aggregate Mining, and Reclamation' (Arbogast *et al.* 2002).[5] This study was based on an inventory and historical review of the extraction and reclamation of infrastructure resources in Metro Denver's Front Range foothills, particularly along Clear Creek, which runs through the city of Golden. The objectives were to assess the compatibility of various, often conflicting, land uses with respect to aggregate mining; to compare the establishment of mining-reclamation landforms at the local level with the regional land-planning vision; and to forecast the future development of other mines downstream by examining the impact of past mining and the process of change (see Figures 6.5, 6.5A, 6.5B).

Figure 6.5 (Refer to note 5 for website information.)

57

Figure 6.5A Aerial photograph, *ca.* 1900, of riparian habitat and braided stream (Clear Creek, Colorado, looking west toward Table Top Mountain).

Figure 6.5B Similar view, *ca.* 2000, with reclaimed sand and gravel pits as water reservoirs, wildlife parks, and commercial land use.

The study considered the geologic setting, a historical perspective, development traditions, ecology, and natural-resources policy along lower Clear Creek (defined for the report as the land between the city of Golden and Clear Creek's confluence with the South Platte River). Multiple sources of information were used to reconstruct the evolution of the landscape along Clear Creek, including journal publications, maps, photographs (terrestrial and aerial), newspaper articles, interviews, and fieldwork. Existing data from a combination of industry, state, and federal reports were used to determine the amount and type of land use in the study area.

The Clear Creek landscape was examined in terms of its culture (e.g. transportation networks, agriculture, and population growth); its environment (stream flow, water quality, pollutant loadings, prairie vegetation, fish and birds, mammals); its natural resources (location and quality of natural aggregate, economic value); and its place (local history of mining and site-specific analyses of selected sand and gravel locations). Tables and figures made use of data from the State of Colorado, U.S. Census Bureau, U.S. Geological Survey Water-Data Reports, and other sources.

The maps and pamphlet illustrate a chronology of human settlement in this corridor. They appraise a specific mineral resource and assess landscape change over approximately one hundred years by using a series of compositions, both in the macro view (the spatial context of urban structure and highways from aerial photographs) and the micro view (the civic scale, where landscape features such as trees, buildings, and sidewalks are included). To merge, identify, and compare attributes more easily, the data were put into a GIS database, using ER Mapper and ArcView software. The raster files were draped across a DEM to provide a visualization of the terrain. The study identified four types of reclaimed mine pits in various locations along Clear Creek: "hidden scenery," water-storage facilities, wildlife/greenbelt space, and multiple-purpose reservoirs. Hidden scenery involves the infilling of pits and quarries and the covering of sites with light industrial, residential housing, or simple vegetation, so that the landform becomes difficult to recognize as a past mine site. The maps of Clear Creek's hidden scenery that were designed for the *Evolution of the Landscape* study won first place in the Blue Pencil competition conducted by the National Association of Government Communicators in 2003.

Thus, reclamation and resource planning can be assessed in science and design terms even if the process is complex and imperfect. The research on the Clear Creek corridor continues today with a focus on investigating and refining descriptive and quantitative models of landscape assessment for use in mining reclamation. The goal is a visual management system that combines aesthetic-quality indicators with landscape-feature inventories relating to the physical and biological aspects of the environment (Arbogast 2003: 24–26).

Summary

In order for form to be as important as function and for meaning to be as important as purpose in the post-mining landscape, landscape architects must consider earth-science issues. This approach brings a richer, more systematic, and comprehensive design to mining reclamation. Landscape architects, along with the general public, must become educated about the environmental impacts of mining. Extensive research, interviews, fieldwork, and data sifting are nothing new in landscape assessment. In order to be an equal partner in the planning process, landscape architects may have to prove themselves as researchers before they are accepted as part of a scientific team. The evolution of one scientist's research and publications as part of the U.S. Geological Survey is a path that began with pure science and has integrated design a little at a time.

This scientist's desire is that people from a design discipline incorporate a scientific background into their ongoing education, thereby better equipping them to create landscapes that are sustainable and respectful of spirit of place, while responding to environmental concerns. By combining science with design, one can use the latter to teach the former. As a group, landscape architects must be multidisciplinary and integrate a variety of purposes. In this manner, the landscape architects who work toward creating complex, healthy landscapes become invaluable to governmental agencies, the extractive industries, and local communities, while garnering respect from fellow scientists.

Disclaimer

Mention of trade names or commercial products does not imply endorsement by the U.S. Geological Survey.

Notes

1 Up-to-date information about science literacy in the United States is available at the National Science Board's website.
2 Up-to-date information about the Canadian Society of Landscape Architecture's position on the relationship between design and science is available at the Society's website.
3 Up-to-date information about the U.S. Geological Survey's objectives is available at the Survey's website.
4 Up-to-date information about the report is available at the U.S. Geological Survey's website (http://greenwood.cr.usgs.gov/pub/circulars/c1191).
5 Up-to-date information about the Clear Creek Corridor reclamation is available at the U.S. Geological Survey's website.

Bibliography

Arbogast, B. (2001) 'A comparison of landscape form and change from anthropogenic and natural earth movement – understanding the public perception,' in R.L. Bon, R.F. Riordan, B.T. Tripp, and S.T. Krukowski (eds) *Proceedings of the 35th Forum on the Geology of Industrial Minerals – The Intermountain West Forum 1999*, Salt Lake City, UT: Utah Geological Survey, 237.
—— (2002) 'An overview of reclamation law dealing with mineral resource development in the USA,' in P.W. Scott and C.M. Bristow (eds) *Industrial Minerals and Extractive Industry Geology*, London: The Geological Society of London, 283–93.
—— (2003) 'A picture is worth a thousand words – reflections on methods to assess the visual impacts of aggregate operations on landscapes,' *Aggregates Manager* 8, 8: 24–26.
Arbogast, B.F., Knepper, D.H., and Langer, W.H. (2000) 'The Human Factor in Mining Reclamation,' *USGS Circular* 1191, Denver: 13.
Arbogast, B., Knepper, D.H., Melick, R.A., and Hickman, J. (2002) 'Evolution of the Landscape along the Clear Creek Corridor, Colorado – Urbanization, Aggregate Mining, and Reclamation,' *USGS Geologic Investigations Series* I-2760, Denver.
Canadian Society of Landscape Architecture (1996) *The Practice of Landscape Architecture*.
Cownover, B. (2004) *Oral communication* (May 2004).
Dempsey, H.S., Todd, J.W., Ferguson, D.L., and Rees, D. (1979) 'The hardrock minerals industry and the landscape architect,' *Environmental Geochemistry and Health* 1, 1: 3, 36–38.
Faigman, D. (1999) *Legal Alchemy: The Use and Misuse of Science in the Law*, New York: W.H. Freeman.
Findlay, C. (1997) 'National geological surveys and the winds of change,' *Nature and Resources* 33, 1: 18–25.
Mayor, F. (2001) Preface, in E. Strosberg (ed.) *Art and Science*, New York: Abbeville Press Publishers.
National Science Board (2000) *Science and Engineering Indicators 2000*, vol. 1, NSB 001, Chapter 8, Arlington, VA: National Science Foundation.
U.S. Geological Survey (2003) 'FY-2003 USGS Annual Science Plan for Geology.'

Part II
Interdisciplinary responses and opportunities in reclamation

Chapter 7
Science, art, and environmental reclamation:
three projects and a few thoughts

T. Allan Comp

Environmental reclamation, and the public support essential to that effort, challenge all of us engaged in the field to broaden our constituency and offer the opportunity to reconnect the sciences and the arts. Drawing on three examples – one in western Pennsylvania, one in southwestern Virginia, and a third on the Oregon coast – this chapter suggests that good reclamation science alone may not be sufficient to address the larger problem. Particularly in places with significant public access or visibility, we need to address the underlying culture as much as the science. Engaging what academia traditionally defines as the arts – history, design, the perspectives of literature, and much more – can significantly improve the public understanding of, and commitment to, our work in environmental improvement; and it can enhance the nature of that response as well. Good design is more than a nice plan; it is also an opportunity for public engagement, even delight. Good history opens opportunities for better understanding of the origins of the contamination we seek to remediate, for greater reflection on the values and achievements of our predecessors and on our contemporary role in that continuum of history and environmental commitment. Equally important, engaging the arts and the sciences as necessary but not sufficient partners brings the depth and range of perspectives essential for expanding community interest in addressing critical needs in our national environmental patrimony.

When one works for a very long time in environmental issues, particularly in reclamation, one begins to realize that environmental problems are created and defined not by science, but by our culture. We as a society decide what we care – and do not care – about at any given time. We define problems as those issues we now care about, and we immediately set about measuring that problem and the potential of its fix, somehow leaping from a culturally defined problem to a scientifically defined fix. While all this measurement, this "science," is necessary, it is not sufficient to fully address the real landscape in which we live. We inherit the sum total of all the previous cultural decisions made about this landscape, and we address those we choose to address. That decision is a cultural decision, but too often we neglect the cultural side of the solution: the arts. I suggest that the vast array of environmental-reclamation science and technology is not sufficient, that the degraded environments we address are cultural artifacts as much as they are problems for science, and that we must address these problems with the full range of the arts and humanities, as well as the sciences, if we are to be effective.

This chapter must also start with a brief caveat. I am, first, a historian of technology and American economics. What follows originated with a dozen years in historic preservation, working on historic industrial districts that were cultural artifacts of great meaning and power, often derelict and inviting new uses. I spent another decade as a developer of historic properties, a consultant to community groups, and on regional heritage development projects. Most recently, I have been working primarily in the Appalachian coal country, perhaps the nation's largest and most damaged cultural landscape, or ecosystem. Coal country and its hundreds of rural communities finally forced me to confront the cultural realities – the attitudes and actions of local residents and their relationship to their environment – that are a critical component of any reclamation effort in this complex landscape.[1]

I focus primarily on acid mine drainage (AMD) remediation in the eastern Appalachian coal country, where the orange streams, coated with AMD, create an emblematic problem of multiple dimensions. I begin this chapter with my first exploration of this challenge to more broadly engage the arts in reclamation, an effort that grew slowly but steadily into something now called AMD&ART, a project located in a small coal camp in western Pennsylvania. Next, I briefly cover a direct outgrowth of AMD&ART, a second project, in southwestern Virginia. A third project, on the Oregon coast, has provided an opportunity to test the same principles (of multidisciplinary collaboration and creative partnering among the sciences and the arts) in a totally different but equally challenging environment.

The AMD&ART Project

The U.S. Environmental Protection Agency recognizes AMD as the largest water-quality problem in the Appalachian coal country, which spreads from northeastern Pennsylvania to central Alabama.[2] Created by chemical reactions between groundwater and abandoned mine workings, this acidic, metals-laden solution seeps or gushes into streams and leaves a telltale orange coating of oxidized iron. Often devastating entire watersheds, rust-colored streams are the remnants of a proud past of hard work and commitment, an era when our national values paid little attention to environmental consequences. Today, AMD is also a painful indicator of the economic abandonment, environmental neglect, and widespread poverty throughout the region, the emblematic orange a silent signature of dying communities.

The thousands of AMD discharges will be addressed by public funds . . . eventually, but the visceral reality of coal-country landscapes – decimated valleys strewn with waste coal, streams flowing orange, poverty at levels two and three times the national average, housing far below adequate – makes more active and effective effort necessary. These were once proud places that contributed heavily to the rise of this nation on its coal-energy base. Companies brought capital and organization to the vast industrial enterprise called coal mining. Families came from across the globe to inhabit the area's coal camps and to work hard toward achieving the American dream. Towns and villages built by coal companies became functioning communities, supporting generations. When the coal companies left, the towns and villages often lived on, but they slowly declined as employment, opportunity, and public attention ebbed.

For outsiders, coal country can be a frustrating place to work, often characterized by what seems like an appallingly passive acceptance of environmental conditions. AMD is everywhere, an orange legacy that leaves dead streams and lost recreational opportunity in its wake. In addition, while many company towns had running water and toilets, the sewer pipe ran directly to the creek: "straight pipes" and fecal coliform contamination are ubiquitous. Effective solutions to these coal-country problems are not easy, and they are not simple. Impoverished communities seldom have the civic capacity to engage larger bureaucracies successfully, and single-discipline solutions fail to gain the understanding that is critical for acquiring public support and funding. There are three or four million people living in coal country, a number too large to ignore and abandon, especially after we, as a nation, made a mess of where they live in our unrestrained quest for energy during the nineteenth and early twentieth centuries. For a Westerner, coming from what Wallace Stegner called the "geography of hope," the eastern Appalachian coal country forces either denial or engagement. I choose the latter.

My epiphany was Site 14 in Somerset County, Pennsylvania, a small AMD-treatment system designed by hydrogeologist Robert Deason. What I saw there was an AMD discharge that was pumped first through an aerator and then flowed through a series of treatment cells (they looked like ponds), which successfully transformed AMD into "legal water" that could be discharged back into the local trout stream. What I also saw, and Bob did not, was a remarkable fountain, surrounded by a cone of black stone and a circular pond of deep red, which then flowed down a tumbling limestone waterway into a series of ponds that, in their increasing biodiversity, clearly reflected the increasing health of the water. When I first remarked on the beauty of the fountain and the interpretive value of this system, I was threatened with bodily harm if I ever again called anything Bob did a "fountain." Happily, he is a teacher of great patience, I am a fast learner, and we soon formed a collegial partnership and mutual interest in a project I was then only thinking about: something I called the "art thing" for a couple of years simply because I did not want to define it any more precisely until I could figure out just what it might be.

What motivated my interest in this "art thing" was the recognition of two realities. First, if AMD were the big problem the EPA said it was, and if I were right in seeing AMD as an emblematic opportunity to address far larger problems of community and environmental health, then any solution would have to engage all of these issues in some way. It would have to bring multiple approaches to the same problem and find ways to integrate those perspectives into a single, holistic approach, one that might serve as a model for others, presuming it worked at all. Second, I could see that this "art thing" was much bigger than science alone: there could be a lot more content in an AMD-remediation site than just good science. It could also be designed in a way that would engage the environmental degradation and abandonment that pervades every coal camp while bringing understanding to both the problem and the solution. It could engage varied interests in many communities, bringing broader support and, with any luck, more diverse funding to the table.

Coal country has a habit of demonizing the other side – labor and management, environmentalists and miners, one ethnic group or class against another – but I realized quickly that there are no clear bad guys in this long history.[3] Without question, the mining techniques and management attitudes this nation and its values once supported have had tragic environmental and social consequences, afflicting thousands of miles of streams and hundreds of communities. In any historical perspective, however, the enemy is ourselves, all of us, as a nation: the values we held at one time created these problems by allowing mining with little thought of reclamation and by allowing company-run towns with little thought of the civic consequences. Fortunately, an assumption embedded deep within the American character dictates that, since "we" did it, "we" ought also to fix it. The challenge for me and the AMD&ART Project was to get enough of "us" to the table and willing to work: no small task in a region so accustomed to environmental devastation that it often goes unnoticed in daily life, a region where the remnant "company town" mind-set makes civic activism a rare occurrence.

A larger perspective is also relevant. The Appalachian coal country is one of America's forgotten places and perhaps its largest forgotten ecosystem. To me, this eastern mountain ecosystem seemed to be a place where the nation might confront – and even overcome – its past environmental and economic values by adding thousands of acres of reclaimed, healthy lands and waters – and peoples – to the national treasure. I thought it might be possible to demonstrate the utility, even necessity, of the arts and the humanities to that recovery process as well, bringing new perspectives, new disciplines, and new supporters to AMD remediation and regional recovery.

Like most of coal country, the townspeople of Vintondale are still deeply connected, in complex and conflicted ways, to the once-central place of coal mining in community life. For the project to succeed, that community engagement had to become an integral component of the AMD&ART process. We could move no faster than the community was willing to move, which was a critical, if occasionally frustrating, necessity. We spent months in the community before we held a first public meeting. When we finally did so, we were pleased to have more than eighty residents (10 percent of the population) turn out for the first two design meetings; they wrote down their ideas on site maps and talked with AMD&ART team members and one another about their town and the best AMD&ART solution to its environmental problems. The resulting design proposal incorporated ideas from everyone who participated, representing the range of interests in the project that was critical to attracting community and agency support, also giving the resulting design the imprint of the community's perspective.

We spent long months being too environmental for the arts funders and too artsy for the environmental funders, but eventually it was the community support and the response of governmental agencies to that support that made the "AMD&ART Park" a reality. I assembled an interdisciplinary design team – a historian, a hydrogeologist, a sculptor, and a landscape architect – and I carefully choreographed the process to ensure that we incorporated the ideas and suggestions of the community and many other sources of expertise around us: state and local agencies, nonprofit watershed groups, or anyone willing to lend an ear or help.[4] It took almost two years from initial exploration to concept plan, but once we had a drawing and the community's approval of what the design should be and what it should include, financial support began to materialize. As a matter of policy, I never made a promise we could not keep, so we never announced a new step in the project until we had the money in hand. Slowly and persistently, I worked with the rest of the design team and our AMD&ART staff of one or two AmeriCorps volunteers to engage the community in transforming this immense swath of desolate land into a place for recreation, historical reflection, ecological education, *and* AMD remediation.

Vintondale, Pennsylvania, is located in the western part of the state, in northern Cambria County. It is an area known for its coal and for the early steel industries in nearby Johnstown, but now it is reduced to less than a quarter of its former population and is crippled by poverty. The AMD&ART site is situated on thirty-five acres of mine-scarred land that was the heart of this small town. The Vinton Coal Company – with its mines, the coal-processing plant with its half dozen major buildings, and the Pennsylvania Railroad line that connected Vintondale with the outside world – once defined this place and its life. The northern edge of the site is the old railroad right-of-way, known today as the Ghost Town Rail Trail, which attracts approximately seventy-five thousand users annually, a major factor in selecting this site. The South Branch of Blacklick Creek, a river severely impacted by AMD, curves around the eastern and southern boundaries of the park and separates it from the town (see Figure 7.1).

At the eastern end of the park, a sequence of wetland treatment cells – shaped to fit the topography and reduce excavation costs – marks the beginning of the treatment system. The AMD discharge runs through this series of settling ponds and a vertical flow pond until it streams into the new wetlands, cleansed of its metallic pollutants and neutralized to a healthy pH. Planted bands of native trees, whose fall colors reflect the increasing health of the water, transition from deep red to orange, yellow, and then silver-green alongside the system; this Litmus Garden is a native-tree arboretum that also creates the opportunity for a fall festival celebrating the Litmus Garden's peak color and Vintondale's recovery.

At the western end of the park, where black bony once barely supported scrubby grasses and stunted trees, a new seven-acre wetland environment, co-designed with the Wildlife Habitat Council, is attracting a variety of birds and wildlife. Excavation for that environment, the project's History Wetlands, revealed the foundation remains of the Vinton Colliery structures, bringing the site's history back to the surface, opening interpretive opportunities.

Finally, at the center of the AMD&ART Park, there is an active recreation area, a place filled with baseball, soccer, horseshoes, volleyball, and grassy places where young children can play safely while their mothers watch from nearby picnic tables. Once the industrial center of old Vintondale and later an environmental liability, the AMD&ART Park is now a community asset; it is fast becoming the educational, recreational, and social hub of the town.

With the overall design artfully developed and painstakingly funded and executed over several years, we could have stopped there. Instead, we shifted our focus to a few unanticipated opportunities to enhance the design and its interpretive values. At the site of the caved-in portal to

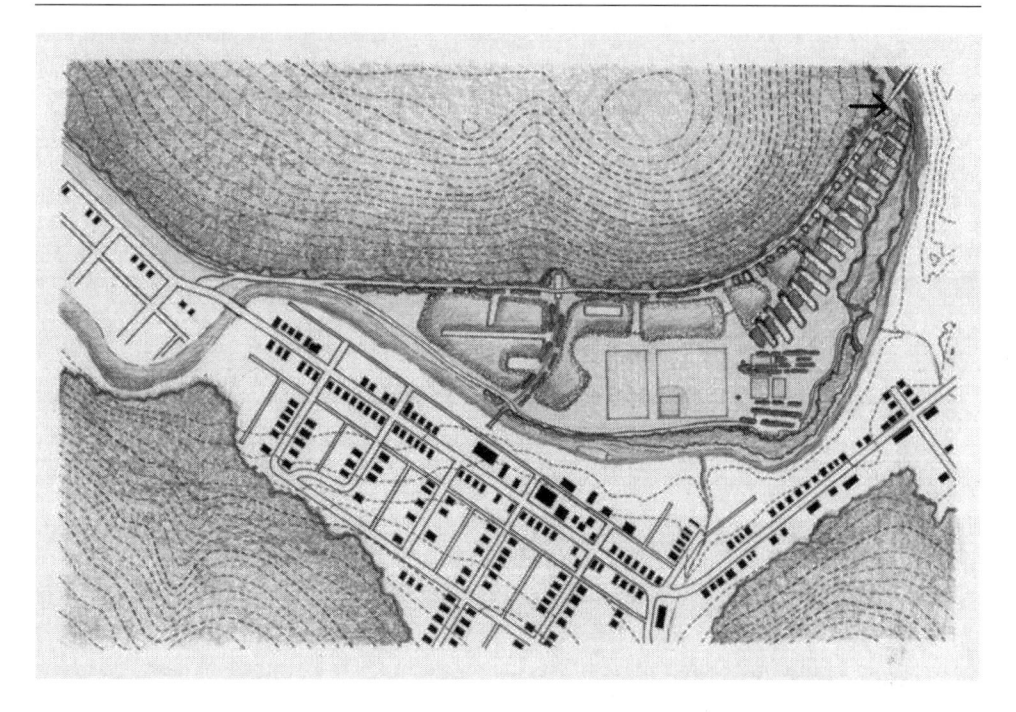

Figure 7.1 AMD&ART site plan, Vintondale, Pennsylvania.

Mine #6, immediately adjacent to the Ghost Town Rail Trail and centered at the edge of the History Wetlands, we reconstructed the heavy timber frame of the original mine portal in its original location and to full scale. We then filled the opening with a polished black slab, hand-etched with the life-size images of miners, taken from some 1938 home-movie footage of men changing shift at this very place, Mine #6. This mural etched in stone is a powerful and immediate reminder of the underground life and its daily threat (see Figure 7.2).

Just across the Ghost Town Rail Trail from the #6 portal, a fifteen-by-twenty-five-foot platform, at grade, has become the interpretive guide to both the surface features of the site and the community just across the river, to work and life above ground. Drawing its source from the original 1923 Sanborn Insurance map of the site, the platform is now a nine-by-fifteen-foot tile mosaic, surrounded by etched images on black tile. These images include company buildings and other community buildings – some now gone – newspaper headlines, and census data. Across the top, we translated the word for "hope" into the twenty-six languages spoken in the community during its heyday between 1910 and 1920. This artwork reconnects mining to community life and supports a better understanding of the aspirations of the many who toiled here, in the mines and at home.

Where the now-clean water returns to the river, AMD&ART hosted a national student competition to create a marker for this community victory over environmental degradation and neglect. The winners exemplified the AMD&ART multidisciplinary approach: an English major and a geology major, both graduate students in landscape architecture at the University of Pennsylvania at the time, won the contest and saw their entry constructed.

All of these later installations have proven remarkably popular with local residents and visitors alike, offering a historical perspective and a more human, individual reference point to what can seem an overwhelming space; it is nearly half a mile long if you walk the Ghost Town Rail Trail. I hope that visitors are encouraged to think a bit about the original scale of this human endeavor;

Figure 7.2 AMD&ART portal #6 full-scale etching, Vintondale, Pennsylvania.

the determination and grit of the people who started it all; and the remarkable commitment of their offspring, two and three generations later, who finished the job with the AMD&ART Park, closing a circle of critical cultural and environmental significance.

Today, visitors to the park experience a place designed by historians, hydrogeologists, sculptors, landscape designers, and many community members, yet never realize that they are in a formally designed place, in which the arts, too, were a full partner in the overall conception of the site. Visitors can walk on interpretive trails that draw together historical information, the science behind passive AMD treatment, and the healing ecosystem that now thrives in the wake of remediation. I hope that residents and visitors alike will gain new perspectives on the resilience of nature and the ability of humans to work with the environment to create a new community center. The physical presence of this energized place can symbolize the success of local residents in healing these waters and this whole site, not only by finishing a job unknowingly abandoned by past generations, but also by developing a new community asset for their families and their families' futures.

It is also worth observing that, while it was the community we sought to benefit and to engage, it was the local, state, and federal agencies that made the project possible. AmeriCorps and Volunteers in Service to America (VISTA) staffed the entire project: there were no paid staff, and there was no paid director; my work with the site was voluntary. Using the wetlands as one example, we were able to engage a complex federal/state permitting process to remove seventy thousand tons of waste material at no cost, saving hundreds of thousands of dollars. We worked with the Army Corps of Engineers to bring lake dredge and other materials for artificial soil, and they provided enough to cover the entire site – wetlands, treatment system, soccer and ball fields – with at least six inches of soil, also at no cost. The Pennsylvania Department of

Transportation (PennDOT) purchased the wetlands as replacement wetlands. AMD&ART then used part of the sale proceeds to construct the wetlands; the rest was used to create an endowment for the treatment system and wetlands at the Community Foundation for the Alleghenies, a fund that will spin off interest annually for maintenance costs. Finally, a huge array of small funders were tapped, some for as little as five hundred dollars, to enable the assembly of artists, scientists, staff, and community members in support of the site that exists today.

This idea of creating a community asset has other, more significant extensions as well. The seeming impossibility of doing anything as big and unprecedented as the AMD&ART Project in an impoverished patch town has been proven wrong. Funding sources as big and as diverse as the Environmental Protection Agency, PennDOT, the Army Corps of Engineers, and the Department of the Interior all brought resources to the table. Private foundations, including Heinz and Rockefeller, also added to the mix, as did the Pennsylvania Council for the Arts and the Pennsylvania Council for the Humanities and many smaller nonprofit foundations. The full listing of those people and organizations who brought their resources and their expertise to Vintondale is on the AMD&ART website; so is the text for every grant we ever received. These agencies and funders supported the AMD&ART Project, I think, because we had every affected community – the town itself, the regional community of AMD experts, the county social-service concerns, the many and varied interests within the town, and many others – as part of the AMD&ART team, comprising the networks of support that are critical for real sustainability.

The UVA-Wise Legacy Wetlands Project

Working on AMD&ART over twelve years as a volunteer director took a significant commitment, but it was also a wonderful opportunity to pursue an idea and all its possibilities – as long as I could raise the money, coordinate the professionals and the many agencies involved, and supervise the AmeriCorps and VISTA staff. Toward the end of the work on AMD&ART, I received a call from a colleague in southwestern Virginia, about five hundred miles and three states south of Vintondale. He offered the chance to test the AMD&ART idea in a different coal-country context. This time the community was a single college campus of two thousand students, and the challenge was to solve a water-quality problem while carrying a message about AMD and the possibility of regional recovery to future generations of students at that campus. The University of Virginia at Wise wanted an AMD treatment system to clean a discharge flowing into a campus lake. My colleague, local Guest River Watershed Coordinator Toby Edwards, and I offered to organize and direct a team that would fix the problem by bringing to the table multiple arts-and-science disciplines, represented by university faculty and local agencies, to create a regional model right in the center of the 367-acre campus of UVA-Wise.

Coal mining in southwestern Virginia has a long history, which is reflected in the landscape and poverty of the region but is not much in evidence at the campus. Early house-coal mining for heating the home of the landowner had left a few AMD seeps in the side of a hill near the chancellor's home, and early twentieth-century strip-mining actually created many of the flat spots on campus that are now occupied by buildings and playing fields. A rapidly expanding branch of the University of Virginia, the new buildings on the Wise campus are starkly modern and the students, meanwhile, are the best the region has to offer. The campus recently constructed a large lake in the center of campus, which serves as both a storm-water catchment basin and a major university focal point. The largest consistent drainage into that lake was an AMD discharge from an old strip mine at the back of the campus, in an area still unreclaimed and undeveloped. Needless to say, the drainage produced significant visual and water-quality problems in this campus showcase. The UVA-Wise Legacy Wetlands Project presented the opportunity to work with a single owner, the university, which demonstrated a major commitment to its community and educational mission. It also presented the challenge of bringing new partners to an academic environment, which seldom engages in either community or outside-agency interactions.

Toby Edwards and I then worked up a proposal for the Wise campus's vice provost. While we agreed that improving water quality, especially in the highly visible lake, was an admirable goal, we suggested that the site had a cultural significance and a serious educational role that an AMD treatment system, done well, could address. In the public eye, this campus already represents self-improvement; it is already a symbol of hope and positive change in the region. The students often come from mining communities, are the first generation of college students in their families, and will likely be the future leaders of the region. An AMD-remediation site in the center of campus, addressing what most presume to be the immutable reality of orange creeks and a depressed economy, could provoke and inform a discussion about the possibility of healing mine-scarred lands *and* communities; it could provide a continuing reminder to generations of students that AMD and other coal-country challenges can be overcome through local creativity and determination.

Working with Kathy Poole, a landscape architect, Toby and I developed for the site a sequence of four wetland-treatment cells, or ponds, which figured both an environmental and an intellectual progression. At the eastern, or upper, end of the site, nearest the chancellor's home, the first cell provides primary AMD treatment, as well as an interpretive opportunity to address what AMD is and what can be done about it (see Figure 7.3). The next cell utilizes the natural topography to form a small island in the center of a wetland pond; a bridge connects this observation platform to a trail that runs throughout the site. The third pond is perhaps the most "natural," with its ragged edges and surrounding riparian-zone vegetation, ascending from wetland to upland forest, a place of numerous opportunities for environmental interpretation and understanding (see Figures 7.4 and 7.5). The last pond is not only more formal but also more widely exposed to casual traffic, offering the final attractive results of reclamation to public appreciation.

When the old AMD discharge flows out of this treatment system and into the upper end of the centerpiece lake, the water is at last clean. Any visitor who takes the same journey can understand a great deal – through site design, signage, plantings of native species, and much more. Through artful design and restrained signage, we address regional economic development and the healing of landscape scars and local economies. We address regional ecology, creating places where students can monitor and do real-world site work. Ultimately, we view the college as a multifaceted educational enterprise, where there are many opportunities and partners to address regional challenges. These messages, carried home by generations of

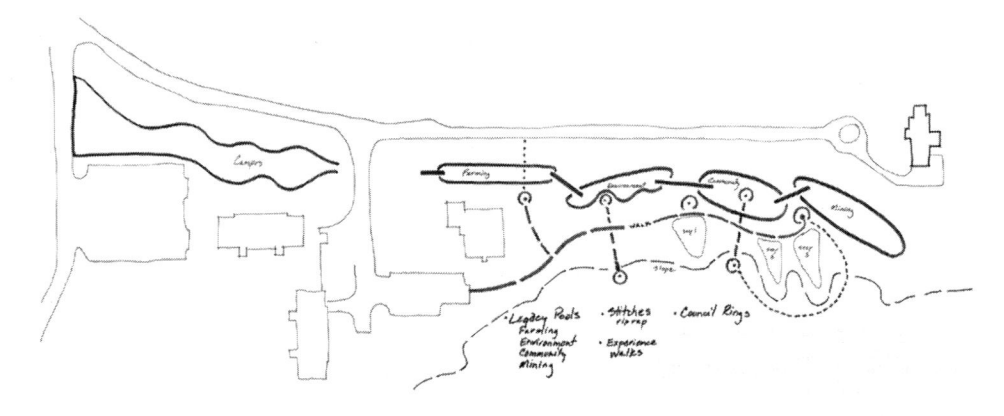

Figure 7.3 University of Virginia-Wise site plan water diagram.

Science, art, and environmental reclamation

Figure 7.4 University of Virginia-Wise site concept model.

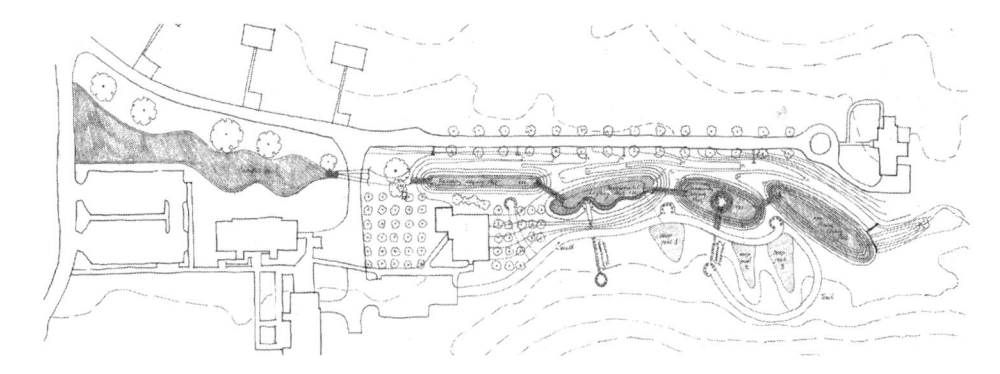

Figure 7.5 University of Virginia-Wise site plan concept.

graduates when they return to their home communities, can initiate a different future for coal country.

It is important to recognize that Vintondale and Wise were not impossible projects; they are, in fact, models, in very different contexts, for the artful assembly of multiple funders and multiple interests. At Wise, it was a nonprofit watershed group, working with a local institution, that needed to solve an aesthetic problem and opened the door to accomplishing much more, including a wide range of educational opportunities for on-campus students, as well as younger visiting students, and even an annual research symposium on the site and its perspectives. Once there was a willing client in the university, it took a team of agencies, among them the Office of Surface Mining (again), the Virginia Department of Transportation, local AMD experts, arts interests, science faculty, and a good designer, whom we hired with some OSM/NEA funding.[5] The project required significant coordination; but the money went to the site, the project is substantially completed, and it is now a growing and evolving presence in the curricula of various departments on campus. It is also a relief to point out that, while the AMD&ART Project took twelve years to complete, the Wise Legacy Wetlands Project took only two. There is some advantage to having some idea of where you are going!

More important, Vintondale and Wise have become catalogs of ideas for AMD remediation in coal country and beyond. They are highly visible places where we have involved the people of coal country in their own history and in their own future. If we design to engage the public, if we interpret remediation sites in ways that encourage understanding, if we make sure that our science is good and the water comes out clean, and if we admit that no single discipline is sufficient, we can create the opportunity for even broader public engagement in environmental recovery. We may even do some small thing to reestablish civic engagement and justifiable civic pride among the communities in coal country.

The Crowley Creek Collaboration ("CCC")

This third project is included not for its grand significance, but for its small assertion that the idea of AMD&ART – that is, that the arts and the sciences belong, *together*, in reclamation – actually works, even in a region radically different from the Appalachian coal country. In fact, places of environmental degradation can be deceptively beautiful, even as the underlying social and economic distress is remarkably severe. Developing a holistic and collaborative restoration plan for Crowley Creek, in Cascade Head, Oregon, was an opportunity to apply what I had learned in Vintondale and Wise to a different historical, ecological, and social context, while maintaining the same, strong commitment to coordinating art, ecology, and the community.

Crowley Creek is the first tributary up the Salmon River on the north-central Oregon coast, an area settled primarily in the twentieth century and primarily by resource-extractive industries (fishing and logging, some farming). If one substitutes timber for coal, the more recent history of these extractive landscapes unfolds in disappointingly familiar ways: economic abandonment, environmental degradation, and the stigmatization of "environmentalism" as a threat to the traditional way of life. On the Oregon coast, the added conflict arises between those who have lived there for generations and those who have arrived only recently to second homes and tourist attractions. To me, Crowley Creek was an ideal reclamation endeavor: the emblematic ordinary place that, on examination, reveals all of the complex cultural challenges of environmental recovery. I proposed to undertake a long-term, intermittent artist residency at the local Sitka Center for Art and Ecology, a nonprofit teaching organization that sponsors artist's residencies, which would allow me to direct an examination of every aspect of the small watershed: its history and ecology, the demands on the water and the land, the interest of the local community in environmental learning opportunities, the interests and concerns of the various government agencies, and the interests and concerns of the local residents, many of whom are also watershed experts of one form or another.

Howard McKee, a local landowner, had then recently granted an educational easement on the only small meadow in the Crowley Creek watershed to the Sitka Center. It was the only open land in the watershed, a meadow of about twenty acres, and the easement provided new and unexamined program possibilities. The Sitka board was committed to both the "Art" and the "Ecology" in the center's name, but the ecology side had been underdeveloped. Moreover, Sitka had not created the ties with the local community and educational agencies that would be critical to successful outreach. The Crowley Creek Collaboration (CCC), as I called the project, therefore set out to explore the potential relationships between the "art" and "ecology" of Sitka's full name, as well as any linkages to surrounding communities (see Figure 7.6).

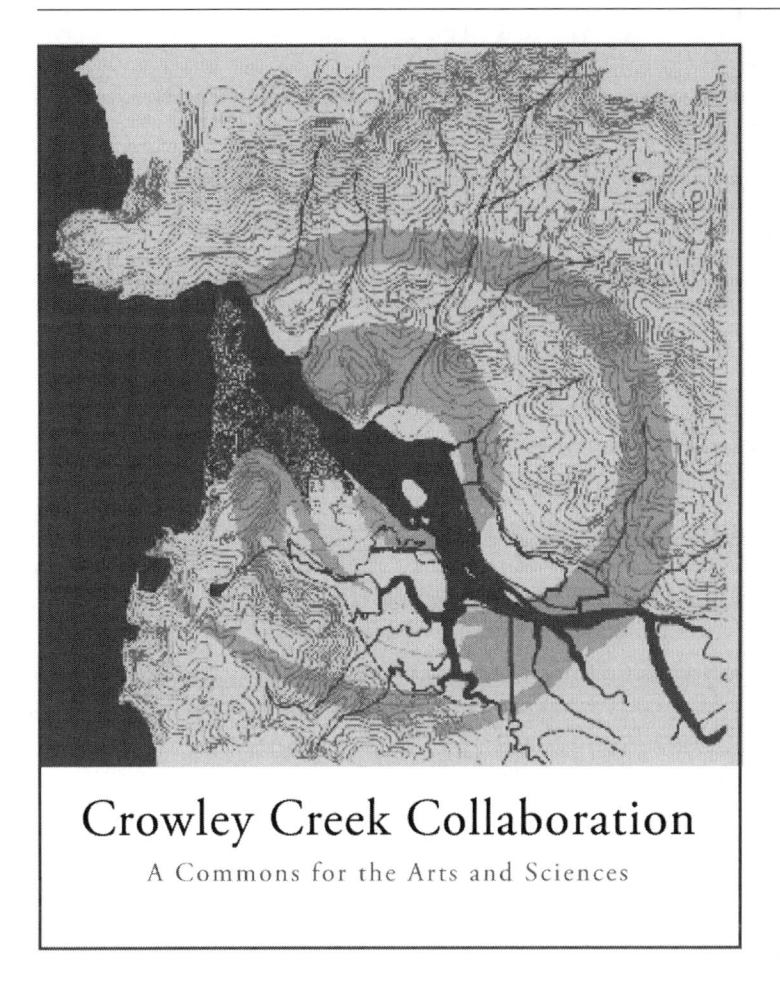

Figure 7.6 Crowley Creek Collaboration booklet cover and location map.

Some initial research quickly revealed the complex web of social, economic, and environmental issues to be faced at Crowley Creek, indeed along the entire Pacific Northwest coast. In short, Crowley Creek stretches from ridgetop to saltwater estuary along a very short distance; it is the last small tributary of the Salmon River before the Pacific, and it is the first stream available to spawning salmon. A myriad of governmental regulatory layers are involved in any action on this land: there is a Federal Research Reserve; the U.S. Forest Service has lands at the top of the watershed and in the Salmon River estuary; and two counties literally divide the meadow. Farming on that meadow and on the rest of the watershed had created serious problems; the creek itself, for example, was basically in a ditch, and water intakes upstream robbed the lower stream of storm-driven gravel flows that are critical to replenishing the spawning habitat. As it happened, many able people were already working on parts of the problem, and they were willing to participate in an unusual planning process that would involve disciplines such as art, design, and history that were new to watershed restoration in the area.

My plan had two basic parts. The first was to develop a realistic proposal for the land by meeting the many layers of regulation and aspiration: restoration of the salmon habitat, restoration of the meadow and its elk habitat, preservation of the existing community water supply, and the facilitation of educational opportunities without inundating the site with wandering students. With the assistance of three VISTA summer positions and some early and adventuresome foundation funding, I was able to direct a lot of basic site research, to find out as much as possible about this ordinary, small place – about the use of the land, the people that live here, the buildings once on the site. I then provided that information to a team of project residents, representing a total of fifteen disciplines, who had agreed to join me in this planning process.

I asked each of the project residents to avoid prescriptions in their individual reports and, instead, to focus on the values and opportunities they saw in this small watershed. Each had the chance to spend a few days on his or her own in the watershed, walking it from top to bottom, and then to prepare a report, which was later distributed to the other team members. My plan was NOT to have a plan until we all met as group. Only then would we start talking about what to do at Crowley Creek as a consensus recommendation from all the disciplines and perspectives that needed to contribute to that conclusion.

It took nearly a year to complete this research phase, but once it was done we were ready to develop a truly multidisciplinary plan, one created by equally informed and equally valued disciplines, working collaboratively. And that's exactly what we did. For nearly three straight days in October of 2005, sixteen of us were in constant conversation from morning through evening. Each of us had read all the reports, each freely brought his or her disciplinary perspective and experience to the table, and none of us was reluctant to speak.

I ended up calling our final recommendations the "CCC Three Rs," drawing on one of the most basic of all learning approaches. Interestingly, our recommendations are remarkably limited in number and scope, primarily because we wanted to work with natural forces, not overcome them. Other reasons for the tight focus of our recommendations are discussed in the conclusion.

The "CCC Three Rs"

We sought to *remove* only what man has put in place that no longer works or that no longer fits with the values we bring to this land in 2005. (In so doing, we concede the temporary nature of our own perspectives and recommendations.) The creek had been managed in the past so that it would stay off the meadow, but it would not be managed as intensively, if at all, in the future. The old culverts, meanwhile, located under an increasingly important access road, worked most of the time, but they would not for much longer. We suggested that we *remove* that old idea of adequate water passage and design a new way for stream and tide to interconnect and flow without restriction, while still providing safe passage for residents.

We would also need to *remember*. Crowley Creek is not a place without a past. We wanted future generations to be reminded of the good lives that earlier inhabitants had lived and enjoyed on this land, from the First Inhabitants to the farmers whose children gathered ferns to sell for candy money. We wanted to *remember* the hard work it took, every time a storm flooded the pasture, to get the stream off the pasture and back into the channel. We wanted to *remember* the generosity inherent in conservation easements, open-space conversions, educational easements, and so much else. And we would need to remember what we are doing now, to consistently monitor and record, photo-document and publish our activities, ensuring that we leave our own record of what we have done for the public memory.

The final need, to *recover*, will follow in our footsteps: we have only to wait. The next big storm or the one after that, the first big tree that falls, the next beaver that comes to stay. . . . All of these will enable the recovery of this place – not back to some mythical "wilderness" state, but toward what it wants to be now and in the future. Crowley Creek will do so because we *remove* past interventions that are no longer appropriate to our contemporary hopes for this place and because we *remember*, with great respect, those who helped create this place and this community, because we remember those who set aside this land for the rest of us. And now we wait, knowing that the environmental drive to *recover* is already underway. In addition, a new partnership between Sitka and the Salmon River Watershed Association, developed during the planning meeting itself, now supports a full-time VISTA position for the project and continues to move forward along the lines developed in the planning report.[6]

Conclusion: a few thoughts on three projects

I think the relative simplicity of the CCC recommendations came about because scientists and humanists and artists – equally informed and equally respected, all necessary, none sufficient – were brought together toward a common goal. With everyone at the table as a participating equal, the final recommendations are more restrained and better balanced. It is also true that, with so many disciplines and perspectives fully represented, no single discipline dominates.

With these three projects now completed and others underway, I think it is clear that restoring watersheds and communities contaminated by AMD, or reclaiming any other environmentally damaged area in which communities have a stake, needs more than a technical fix. Sitka may be deceptively beautiful, but the layers of cultural complexity – in the residents' communities, in the local, state, and federal agencies – create the real challenges. Sustainable reclamation cannot be just a science project! Lasting solutions to the multifaceted problems of environmental reclamation must be cultural *and* environmental. A scientific solution may clean the water or address some other specific problem, but a truly collaborative, multidisciplinary approach that engages the arts and the sciences has the power both to clean the water and involve the community in a healing process that continues long after the water is clean.

In the process of doing these three projects, I have tried to establish a new and different role for artists and humanists as well: not as solitary visionaries, but as participants; not as some ultimately mystical or magical intervention, but offering an important, critical perspective; not as an arbitrator, but as a coworker, one among many disciplines equally necessary to the recovery and revitalization of a region and its peoples. It is not easy to adapt to new roles and new kinds of relationships with other disciplines, but I think it is critical. I know from experience that real collaboration has to be very carefully developed – and then carefully managed to ensure continued cooperation.[7]

I've also realized what a critical role local, state, and federal agencies have played in each of these projects. Without local support and expertise, without state-agency funding and permission, and without significant funding – and the credibility bestowed by the receipt of that funding – from federal agencies, AMD&ART simply would not exist. When we secured the interest of the vice provost at UVA-Wise, we immediately assembled a team of local and state agencies, including my own Department of the Interior Office of Surface Mining and its in-state counterparts, as well as private-sector consultants who agreed to work pro bono. They all participated because they, like everyone else, wanted to do something adventuresome and right. At Sitka, when we started to involve the local watershed agencies, the Nature Conservancy and its constituency, both public and private educational institutions, and the private landowners, then the opportunities for collaboration really started. In each case, it was government agencies (and their local employees), even more than the individual community members, that helped push for innovation, for breadth of vision, and for sustainable results. It was their entrepreneurial willingness to participate that created the best opportunities for collaboration.

Finally, I continue to be impressed by the range and capacities of the agencies and individuals that are willing to come to the table and cooperate in environmental reclamation. They are not all scientists – indeed, the scientists often find them intrusive – but they represent the actual community of interests that must be brought to this endeavor. It is possible to reconnect the sciences and the arts, to bring the whole of our human experience to reclamation, and to establish a significantly wider circle of partnerships for our work. It takes public and private partners;

local, regional, and national perspectives; academic institutions and practitioners; community volunteers: all necessary, none sufficient.

It took twelve years to finish AMD&ART, two to get the UVA-Wise Legacy Wetlands constructed, and just one year to get the Crowley Creek Collaboration through a complex planning process and on its way. Obviously, that sequence demonstrates the great utility of knowing a little about where you are headed, or at least a bit about how to get there. I hope that these examples and their respective websites will encourage others to attempt such projects as well. If we stretch far enough to get really good at this multidisciplinary and collaborative endeavor, if we remember the critical role of multiple partnerships, and if we keep the cultural side of reclamation balanced with the science, we just might create the vital public engagement that fosters better understanding, and even healing, while securing the funding on which this good work so heavily depends.

Notes

1 This chapter is not the official opinion of the Office of Surface Mining, where I still work full-time. It is my personal perspective, based largely on the research and project development work I do on my own. This is the second year I have been asked to write about P-REX. I appreciate the opportunity and intend to take full advantage of it. For complete coverage of the projects I discuss, see the websites for the AMD&ART Project in Vintondale, Pennsylvania (www.amdandart.info), the Crowley Creek Collaboration in Cascade Head, Oregon (www.sitkacenter.org/ccc/index.html), and the UVA-Wise Legacy Wetlands Project in Wise, Virginia (www.uvawise.edu/wetlands_symposium_06/wetlandshistory.html).
2 The core of the Appalachian coal country runs down the spine of the Appalachian Mountains, including the anthracite region of eastern Pennsylvania, most of western Pennsylvania, eastern Ohio, western Maryland, virtually all of West Virginia, southwestern Virginia, eastern Kentucky and Tennessee, and central Alabama.
3 To be sure, there were lots of bad actors on both sides. My point here is only that the blame can be fairly placed on our national values without any need to blame one side or another.
4 The core design team for the Vintondale project: T. Allan Comp, PhD, historian; Robert Deason, hydrogeologist; Stacy Levy, sculptor; Julie Bargmann, landscape designer. For a listing of all the partners, funders, and staff who helped to create the full AMD&ART Project see the "Staff and Board" area of the AMD&ART website (www.amdandart.info/staffboard.html).
5 Kathy Poole, now of Biohabitats, and I were co-designers of the site. We worked with the steering committee of local agencies and the university.
6 The entire Crowley Creek Collaboration report is available at the CCC area of the Sitka Center website (www.sitkacenter.org/cccproposal.html).
7 I would quickly add that real collaboration is not for everyone. Being part of a team means that one does not get exclusive credit for anything, which can make life difficult for those trained in disciplines that are driven by personal recognition and for some young professionals at critical points in their careers.

Chapter 8
The Wellington Oro mine-site cleanup:
integrating the cleanup of an abandoned mine site with the community's vision of land preservation and affordable housing

Victor Ketellapper

Introduction

Historically, in the western United States, mineral mining has been a temporary use of the land. After the economically recoverable minerals had been removed, the mines were abandoned, impacting the characteristics of the property. These changes in the property included waste-rock piles, mineral-recovery processing wastes, and continuous discharges of acidic metal-laden water, known as acid mine drainage. They often result in adverse environmental or human-health impacts.

The number of abandoned mining sites in the United States is unknown. Estimates range from two hundred thousand to over five hundred thousand. The discrepancy is a result of how different sources define a mine site. Fortunately, only a small fraction of these mines cause significant environmental problems. However, a substantial aggregate impact from mining exists. In 1993, the U.S. Forest Service estimated that five to ten thousand miles of streams and rivers within National Forest lands are affected by acid drainage from mines (Office of Water 1997: 2).

Some of the abandoned mineral-mining sites that exhibit significant impacts on human health or the environment are being addressed under the United States Environmental Protection Agency's (EPA's) Superfund Program, which investigates and cleans up hazardous wastes at abandoned industrial facilities. Currently, there are more than sixty mining or mining-related sites designated as Superfund sites. These represent some of the most expensive and time-intensive sites to remediate, with costs in excess of $25 million and taking over 10 years to complete. For example, the cleanup costs at the Summitville Mine Superfund site, a 500-acre abandoned open-pit gold mine in the mountains of Southwest Colorado, are approaching $200 million. Cleanup began in December, 1992 when the site was abandoned by the mining company, and is planned to be completed by 2010. After the mine reclamation is completed, the expenditures to maintain the remediation will be over $2 million per year, with the bulk of the costs for the treatment of the millions of gallons of acid mine drainage generated at the site annually. The majority of this work is and will be funded by the federal and state governments (see Figure 8.1).

The cleanup of environmental problems at abandoned mine sites is a multifaceted combination of technical, financial, and liability challenges. Without successfully addressing every issue, cleanup moves very slowly, if at all. This is certainly true at the French Gulch site, a century-old

Victor Ketellapper

Figure 8.1 Summitville Mine Superfund site.

metals-mining district, located just outside Breckenridge, Colorado: the mine owners and operators either no longer exist or have limited resources, and those wishing to help voluntarily fix the problem have been scared off by laws that would hold them responsible if the cleanup were insufficient or cleanup standards changed.

The impacts from the abandoned mining in French Creek were sufficiently significant that EPA was considering the site to be included in the Superfund Program. However, local government officials resisted this due to their concern over the effect that EPA Superfund involvement would have on their community and tourist economy. A community-based approach was proposed by EPA to resolve these issues. This led to the formation of the French Gulch Remediation Opportunities Group, a stakeholders group formed to develop plans to address mining-related environmental impacts within French Gulch. This common vision provided a foundation for a unique multiparty settlement addressing the situation at the French Gulch site that defined environmental liability, provided funding for mine reclamation projects, and allowed the purchase of abandoned mines and adjacent properties for open space, outdoor recreation, habitat preservation, historical preservation, and affordable housing. The details and results of the successful implementation of this new approach at the French Gulch site are presented in this chapter.

The French Gulch Remediation Opportunities Group (FROG)

In order to promote stakeholder involvement and a collaborative process for environmental decision making at the French Gulch site, the French Gulch Remediation Opportunities Group (FROG) was formed in the spring of 1995. The participants included citizens, county commissioners, Breckenridge town-council members, the Breckenridge town manager, county and city open-space managers, ski-area representatives, landowners, and state and federal regulatory and land-management agency representatives. B&B Mines, the company that owned the Wellington Oro Mine, as well as more than 1,800 acres of mining claims in French Gulch and the adjacent Swan River watershed, also participated in the FROG. The FROG was initially facilitated by the Keystone Center, a nonprofit organization that provides public-policy mediation services.

At the time the FROG was formed, B&B Mines no longer had active operations and therefore had a very limited income and no cash reserves to conduct reclamation activities. The company would therefore need to sell the property in order to finance the reclamation of its historic mining operations. Thus, future land use became an important topic of discussion as the FROG developed plans for reclaiming the mine. At that time, French Gulch served as a backcountry recreational area for the community of Breckenridge. That use would change dramatically if the mining company's one thousand eight hundred acres were developed for vacation homes as experienced throughout Summit County. Moreover, development would inevitably affect the historic mining landscapes and ecological habitats.

EPA provided funding to facilitate the FROG's grass roots efforts in developing an integrated solution to community and environmental issues at French Gulch. The focuses of these early discussions were on environmental liability concerns and developing an understanding of the magnitude and sources of contamination. After years of studying and debating technical and legal issues, in 2001 a common vision that integrated the community's vision for land use in French Gulch and mine reclamation was achieved. The three primary aspects of this vision were: 1) to clean up the abandoned mines that potentially or actually impacted human health or the environment; 2) to build a livable, affordable housing community for locals who work in the resort town of Breckenridge; and 3) to preserve the mining company's one thousand eight hundred acres of land as open space for recreation and wildlife habitat. The development of this vision did not include a landscape architectural design professional. I believe that the services of a landscape architect would have expedited the process through better integration of future land use discussions.

The French Gulch site and mining history

The French Gulch site is a watershed impacted by abandoned mines located in the Rocky Mountains east of the Town of Breckenridge, Colorado. The two primary environmental concerns associated with the district's mine sites are the aquatic-life impacts from acid mine drainage discharging from the abandoned Wellington Oro Mine and the potential human-health impacts from long-term exposure to mine wastes with elevated levels of lead and arsenic.

Extensive placer and underground lode mining occurred in French Gulch from the late 1850s to the 1960s. Placer-gold mining began in French Gulch in 1859 with small gravity-separation operations; the dredging operations that followed continued until the 1940s. The dredging operations resulted in forty to fifty foot high piles and ridges of cobbles and gravel-size placer tailings throughout the valley floor (see Figures 8.2 and 8.3). Underground lode mining began in 1889 and continued through the 1960s. The underground mines typically produced high-grade zinc-lead-silver ores, as well as some gold ores. Underground mining left numerous shafts, adits, waste rock, and tailings throughout the Gulch.

The Wellington Oro Mine and Mill complex was the largest mining operation on the site. Producing lead-zinc-copper-sulfide and gold ores, it operated from the 1880s to the 1930s. The underground mine workings consisted of more than twelve miles of tunnels, adits, drifts, stopes, and crosscuts and extended more than a thousand feet below the ground surface. Major portions of these mine workings are now flooded. As water migrates through the mine, it reacts with the mineralized rock generating sulfuric acid, and the acidic water then draws metals into solution. At the Wellington Oro Mine, the water migrating through the mine becomes contaminated with zinc and cadmium. Eventually, this contaminated water reaches and mixes with water in French Creek, causing French Creek to become toxic to aquatic life (see Figure 8.4).

Figure 8.2 Abandoned dredge in French Gulch used to recover gold from river bed.

Figure 8.3 Acid mine drainage from the Wellington Oro Mine. The Wellington Oro Mine is seen in the background. Piles of rock are the mining spoils from placer mining of French Gulch.

Figure 8.4 Physical aquatic habitat damaged by placer mining in French Gulch. Encroachment of housing prevents restoration to a fully functional habitat.

French Creek water quality and aquatic habitat

The FROG concerned itself with developing goals for water quality and aquatic habitat in the Blue River and French Creek that had been impacted by historic mining. The primary impact to water quality was caused by acid mine drainage generated from underground mining operations at the Wellington Oro Mine. The physical habitat had been destroyed by the extensive dredging operations throughout French Gulch and the Blue River. To restore a functioning aquatic ecosystem, both water quality and habitat would require improvement. In French Gulch, the question became: What aquatic life could be supported by feasible improvements to water quality and habitat?

The water-quality standards established for the French Creek and Blue River by the State of Colorado were those dictated by an EPA Clean Water Act Class 1 Cold Water Fishery classification, designed to protect all cold-water species, including species very sensitive to dissolved metal concentrations. While this standard is commonly applied to all high mountain streams in Colorado, site-specific studies were needed at French Gulch to understand whether the Class 1 Cold Water Fishery water-quality standards could be achieved, or whether a site-specific water-quality standard should be considered that was protective of the species expected in a restored habitat.

Concern over water quality in French Creek was first raised in 1989, when fingerling trout released into the Blue River, downstream of French Creek, died. The Colorado Department of Public Health and the Environment (CDPHE) conducted water-quality sampling in response to this fish kill, and the results showed levels of zinc that were acutely toxic to trout in French Creek below the Wellington Oro Mine and into the Blue River. From 1996 to 1999, the surface-water chemistry of French Creek and the Blue River was monitored by EPA to evaluate sources of metal discharges, the metal loadings (mass of metal release per day) associated with each source, and the impact of in-stream metal concentrations from each source. This data demonstrated that the primary metal load in French Creek originated from the Wellington Oro Mine (URS 2002: 10). It was reaffirmed that elevated levels of dissolved zinc and cadmium in the surface water of French Creek and the Blue River downstream of the Wellington Oro Mine were typically at levels acutely toxic in French Creek and chronically toxic in the Blue River for three miles below its confluence with French Creek to adult fish and invertebrates. The concentrations of these metals were, moreover, at high enough levels to be toxic to trout eggs and fry (SRC 2002: 7–23).

Further studies identified the primary source of surface-water toxicity appeared to be the result of flooding of the underground workings of the Wellington Oro Mine. Investigations revealed that contaminated water from the mine travels from the flooded mine workings through highly fractured bedrock to the alluvium of French Creek and results in a diffuse discharge of acid mine drainage over a large area, so that constructing an effective system for preventing or capturing contaminated water migrating out of the mine pool to French Creek would be difficult if not impossible. Thus, it would be impractical to completely remove this source of contamination from the Blue River and French Creek, making it unlikely that Class 1 Cold Water Fishery water-quality standards could be attained.

At this point, the focus moved from water quality to understanding the placer-mining-disrupted physical habitat. Studies were initiated in 2003 to evaluate the physical habitat to support aquatic life and its potential for restoration. It was found that the lower part of French Creek was highly channelized and development along its banks would prevent restoration of a habitat that would support a trout fishery (see Figure 8.5).

The habitat assessment of the segment of the Blue River that was impacted by metals contamination from French Creek found that a portion of the river had been restored and that plans

Figure 8.5 The cementation of boulders to create the kayak course limits aquatic habitat in the Blue River.

had been developed to restore the remaining portions of this stretch. However, the restoration resulted in an aquatic habitat that did not support spawning. In this segment of the Blue River, a kayak course with the stream bed held in place with concrete had been constructed. This resulted in conditions unsuitable for trout egg survival and benthic communities. During low flow periods, there was no water flowing in parts of the stream, a result of the porous placer mining dredge spoils allowing water to travel underground rather than in the stream bed (see Figure 8.6).

The FROG evaluated this data, coming to the conclusion that a realistic goal for mine reclamation would be an adult brown-trout fishery for the Blue River and the ambient water quality in French Creek. The adult brown-trout fishery goal would allow for slightly higher concentrations of dissolved zinc and cadmium, that could be attained by feasible mine reclamation technologies. In French Creek, restrictions to improving the physical habitat prevented reestablishment of a functioning habitat for aquatic life. Thus, improving water quality to aquatic life standards was found to be unnecessary. These revised water-quality standards were proposed to the Colorado Water Quality Control Commission in 2003 and approved (Wyatt and Vieira 2003: 1–4).

With the site-specific water-quality standards in place, the focus was placed on engineering solutions to improve water quality. The EPA's and CDPHE's investigations had revealed that a significant portion of the metal load in French Creek originated from a single spring, which was fed from a fault that passed through the Wellington Oro Mine (URS 2003: 10). A system could be easily designed to capture and treat this single identified point source discharging from the Wellington Oro Mine. An evaluation was then conducted to predict the water quality that would be achieved if this readily collectable seepage from the Wellington Oro Mine was treated to remove the zinc and cadmium contamination. The evaluation found that water quality would

Figure 8.6 During the summer time, the Blue River often runs dry. This prevents the survival of trout eggs.

improve by treating this seepage with the prospect of achieving the revised water-quality goals. The EPA then conducted an engineering feasibility study to evaluate alternatives for the collection and treatment of the readily collectable seepage. The findings were presented to the public in May 2004. The public was supportive of EPA's proposed cleanup plan that included the collection and treatment of acid mine drainage seeping from the Wellington Oro Mine spring. The estimated cost to implement the plan, including long-term operations, was $5 million.

Affordable housing

People who work in the historic resort town of Breckenridge are being forced to live out of the area by median costs of $725,000 for a single-family home. For many workers, affordable single-family homes were only available across Hoosier Pass, a forty-five-minute commute over treacherous mountain roads.

A developer, David O'Neil, proposed to the FROG an eighty-five-acre site in French Gulch, on the town's outskirts, as an ideal location for the development of an affordable-housing community. This property, however, was impacted by historic placer-mining activities and by the adjacent Union Mine and Mill site. Elevated levels of lead and arsenic in soils on a small portion of the property would be a hazard to human health if the property were developed into a residential community. The developer proposed to construct affordable housing and conduct mine reclamation on the property. The proposal was accepted by the town and the mining company. The community was named the Wellington Neighborhood.

The approval of the subdivision prompted the U.S. Forest Service to reclaim the Union Mine and Mill site that was located on land it managed. The U.S. Forest Service was concerned that the mining wastes bearing elevated levels of lead and arsenic would pose an unnecessary health risk to residents of the Wellington Neighborhood.

To address the environmental liability concerns raised by the residential development, a Prospective Purchaser Agreement was negotiated and signed by the developer, the EPA, and CDPHE. This agreement described the mine reclamation work that the developer agreed to conduct on the property, and established lead and arsenic soil-action levels of five hundred parts per million (ppm) and fifty ppm, respectively. Since the property was being purchased from B&B Mines, the company responsible for a portion of the mine reclamation work in French Gulch, $850,000 generated from the sale of the property was set aside for future investigation, design, and reclamation of mining sites throughout French Gulch.

The first phase of the Wellington Neighborhood affordable-housing development was completed in 2006. Eighty percent of the homes constructed were reserved for purchase by people who work in Summit County at about one-third the cost of the median home-purchase price in Breckenridge. These homes have a deed restriction that limits the appreciation of the resale price of the property to 3 percent per year or the annual percentage increase of the "area medium income," whichever is greater. Homeowners include the town manager, government employees, shop owners, teachers, and police officers. In 2002, the Wellington Neighborhood was recognized by the EPA with a National Smart Growth Award as a model for resort communities where affordable housing for permanent residents is scarce (see Figure 8.7).

Open-space preservation

The FROG raised concerns about the potential of sprawl-type development throughout French Gulch. The construction of large summer homes was occurring throughout Summit County. French Gulch, however, was being used by the residents and tourists for backcountry outdoor recreational activities (such as hiking, mountain biking, off-roading, cross-country skiing, and hunting), all of which would be compromised by summer-home development. French Gulch also contained historic mining landscapes, and in the upper portion of the drainage basin, habitat for a Colorado species of concern, the Colorado River cutthroat trout, that would be

Figure 8.7 The Wellington Neighborhood, an affordable housing community constructed on reclaimed placer-mining spoils.

adversely impacted by such development. The consensus of the FROG was to preserve the remaining 1,800 acres of B&B-owned properties as open space.

In November 2001, through negotiations independent of the FROG, the Open Space Departments of the Town of Breckenridge and Summit County signed a sales agreement with the mining company for the purchase of its more than one thousand eight hundred acres of land in French Gulch and in the adjacent Swan River watershed for $9 million. The properties consisted of 170 mining claims, including the Wellington Oro Mine. As part of the sales agreement, Summit County and the Town of Breckenridge agreed to construct and operate the water-treatment remedy at the Wellington Oro Mine. The buyers also agreed to conduct surface reclamation at an additional two mine sites (see Figure 8.8).

As part of this real estate transaction, two additional agreements were negotiated in order to resolve environmental-liability issues. First, Summit County and the Town of Breckenridge negotiated with the EPA and the State of Colorado a Bonafide Brownfields Agreement. This agreement defines the specific mine-reclamation activities to be conducted by the town and county, and limits their cleanup liability on other abandoned mine sites purchased. The second agreement resolves B&B Mines Superfund environmental liability, including past and future environmental cleanup costs. This agreement allowed the mining company to dissolve and distribute its assets to its stockholders. Although the negotiations among the multiple parties to finalize these agreements were complex, these settlements were successfully completed in March 2006.

In 2006, the Town of Breckenridge and Summit County completed the surface reclamation of the two abandoned mining sites. The construction of the water treatment plant at the Wellington Oro Mine is scheduled to be completed in 2007. Also in 2007, the EPA, the Town of Breckenridge and Summit County will be developing a conceptual plan utilizing the landscape architectural services of the Project for Reclamation Excellence (P-REX). This plan will integrate

Figure 8.8 An overview of French Gulch which is being preserved as open space for habitat preservation and backcountry recreation.

mine reclamation, expansion of aquatic habitats for threatened species, integration of the recreational use plan, and preservation of historic mining artifacts.

Conclusion

The French Gulch site provides an example of how mine reclamation and redevelopment planning can be conducted concurrently. At French Gulch, a collaborative process provided a forum for discussing the varying needs of multiple stakeholders. The collaborative process resulted in a common vision that balances stakeholders' three primary concerns in French Gulch: correcting environmental problems associated with abandoned mining, providing affordable housing, and open-space preservation. This vision led to a common ground among the stakeholders, providing the desire for the parties to resolve difficult liability issues. The resulting settlement provided for the mine reclamation to be conducted without additional funding from EPA's Superfund Program, including the costly construction and operation of a facility to treat contaminated water discharging from the Wellington Oro Mine.

Bibliography

Office of Water, Environmental Protection Agency (EPA) (1997) *EPA's National Hardrock Mining Framework*, Washington D.C.: EPA.
Syracuse Research Corporation (SRC) (2002) *Ecological Risk Assessment for the French Gulch/ Wellington Oro Mine Site*, Denver, CO: Syracuse Research Corporation.
URS Operating Services, Inc. (2003) *Wellington Oro Mine Pool Engineering Evaluation/Cost Analysis, French Gulch Site*, Denver, CO: URS Operating Services, Inc.
Wyatt, L. and Vieira, N. (2003) *Use-Attainability Analysis, Lower French Gulch and the Blue River Downstream of French Gulch near Breckenridge, Summit County, Colorado*, Silvertorne, CO: Summit Water Quality Committee.

Chapter 9
Building partnerships for post-mining regeneration:
Post-Mining Alliance at the Eden Project

Caroline Digby

Throughout the twentieth century, Cornwall, in the far southwest of England, experienced the protracted contraction of its traditional primary industries: fishing, agriculture, and mining. The county's mining history is long and proud: four thousand years of metal extraction peaked in the nineteenth century when the county produced over half the world's copper and tin, and developed and exported the hard-rock-mining industrial revolution around the world. Metal mining declined through the late-nineteenth and twentieth centuries, however, with the last tin mine, South Crofty, closing in 1998. China-clay extraction, which began in the mid-eighteenth century, produces about two million tons of kaolin for export, but as the industry has become more capital-intensive, it has had a series of large retrenchments in the last fifteen years. The resulting socioeconomic decline has been exacerbated by a dispersed rural population, the absence of a large population focus, the industry's distance from large urban centers, by market failures, and by a lack of civic leadership.

Tourism, based on a spectacular coastline, has saved the county from complete economic meltdown, certainly since the mid-twentieth century, and constitutes 19 percent of Cornwall's economy, employing around 16 percent of the workforce. Tourism brings its own social and economic challenges, however, including seasonal employment, low wages and low skills, dependency, social exclusion, and house-price inflation.

Launched in March 2001, the Eden Project has had a transformational effect on the society and economy of Cornwall. Constructed in a 160-year-old, exhausted china-clay quarry at Bodelva, near St Austell, Eden is an educational charity and plant-based visitor attraction that aims to connect people, plants, and natural resources. Since opening in 2001, it has attracted more than eight million people and has drawn more than £750 million into the local economy. In all its policies, projects, and programs, Eden attempts to maximize its economic, social, and environmental benefits; to this end, innovation permeates the organization's civil engineering, construction, materials sourcing, waste management, employment policy, management structure, external relations, educational programs, public learning and interpretation, plant health, nutrition and soils, horticultural and exhibit design, and more (see Figures 9.1A, 9.1B, and 9.2).

A fundamentally unique reclamation project, by taking many of the principles of conventional mine rehabilitation and pushing them to the extreme, Eden has proven that, with some imagination and determination, people can produce an enormous positive force for change. The project symbolizes what can be achieved by delivering an ambitious vision born of the local circumstances of Cornwall's post-industrial decline. Indeed, Eden exemplifies ideally a central rationale behind the philosophy of the Post-Mining Alliance: although it was never conceived as

Figure 9.1A Bodelva Pit 1999.

Figure 9.1B Biomes under construction 2000.

Figure 9.2 The Eden Project 2005.

a post-mining regeneration project per se, that is indeed what it has turned out to be. The challenge for the wider mining industry and related government agencies is how to harness this passion and creativity to drive new solutions to old problems.

Eden bought the Bodelva china-clay quarry in 1998 for £4.2 million. The construction of Eden was a major event, generating huge amounts of public and media interest. An unstable, muddy china-clay pit, forty meters below the water table in one of the wettest parts of the United Kingdom, does not appear at first sight to be the ideal location for creating a multimillion-pound flagship visitor attraction. So why was it chosen? There were three main reasons:

- it would provide a beneficial microclimate for plants;
- almost complete concealment of the site would create a dramatic spectacle at the entrance to the Visitor Center; and
- its success would demonstrate that degraded land can be transformed.

Such a site poses many challenges; a few of those challenges are summarized here:

- the relocation of 1.5 million tons of fill material to create the basic landforms;
- the installation of a state-of-the-art drainage and pumping system;
- the stabilization of the slopes according to the competence of the granite bedrock, requiring hydroseeding, the use of one thousand five hundred rock bolts, and some nine thousand square meters of sprayed concrete;
- construction of the world's biggest and most architecturally challenging greenhouses;
- provision for wheelchair accessibility throughout the whole site; and
- the creation of eighty-five thousand tons of artificial soil from recycled materials.

Construction of the two iconic conservatories, or biomes, began in December 1999. The Humid Tropics Biome (HTB) is the largest greenhouse in the world. It is two hundred meters long, a hundred meters wide, and fifty meters high and is constructed from tessellated polygons in a geodesic-dome design. Consequently, it is self-supporting, it requires no internal supports, and, owing to its revolutionary design, it weighs only four hundred tons – an efficient use of materials for such a huge structure. The Warm Temperate Biome (WTB) is similarly constructed and approximately half the size of its neighbor. For erecting the giant Meccano framework of these domes, Eden received two Guinness World Records for the tallest and largest freestanding scaffolding structure ever. The birdcage scaffold required more than a hundred thousand poles.

The glazing material is a strong, lightweight, self-cleaning plastic called ethyl-tetra-fluoro-ethylene (ETFE), chemically similar to the PTFE coating of a nonstick frying pan. Glass would have rendered the whole design impractical, owing to its weight. The windows consist of triple-glazed pillows, the individual sheets being held apart by compressed air. The HTB alone has more than five hundred of these windows, the largest of which is eleven meters across and voluminous enough to accommodate a London taxi.

The first key building to open to the public – the Visitor Center – opened in May 2000, during the main construction phase. In six months, five hundred thousand people visited to observe the evolving landscape and spectacular skeletal biomes. Since the opening of the whole site on March 17, 2001, several additional phases of construction have completed the Eden landscape with such new structures as the Foundation Building and, in 2005, Eden's educational-resource center, the Core. The buildings aim to push the boundaries of architecture and construction practice, and all strive to showcase innovation and sustainability in design and construction.

As "the living theatre of plants and people," Eden relays its educational mission through approximately ninety plant-based exhibits, aimed particularly at those who have never really considered the importance of plants (or, by extension, related environmental and social issues). Eden's plant collection is an educational collection, consisting primarily of the common plants that are used every day around the world for basic necessities like food, construction, medicines, dyes, textiles, paper, and so on. Its philosophy stands in contrast to the approach of more traditional botanic gardens, which often act as repositories for economically valuable or ecologically interesting species. Bananas, for example, are the most popular fruit in the United Kingdom, yet very few British people have seen bananas growing on banana plants. Eden's mini plantation of bananas allows visitors to see, feel, and smell the fruit – a sensory experience that is then used to explain tropical agriculture, world trade, nutrition, and British connections with the faraway people who grow its food. Similar approaches are used throughout the Eden site to showcase the vital relationship of people, plants, and resources.

Using the plant collection as the educational backdrop, the visitor is engaged emotionally through a diverse range of activities. The use of text-based boards and video screens is minimized in favor of personal interaction (performance, guiding, storytelling) and art-led installations and imagery. This approach is integral to Eden's ethos of encouraging artists, scientists, engineers, horticulturalists, designers, and accessibility experts to collaborate. Outwardly chaotic, it has achieved renown for its refreshing provocation of the public's engagement with complex social and environmental messages, and that the same approach is now being adopted by many other plant-based visitor attractions.

Eden's fifteen-hectare visitor-accessible landscape is currently divided into four main areas. The Humid Tropics Biome (1.2 ha), sometimes known more colorfully as the largest rainforest in captivity, houses plants from the tropical regions of the world, focusing on four geographical areas – Amazonia, West Africa, Malaysia, and the Oceanic Islands – and features a large area, called Cornucopia, that is devoted to the commercialization of the plant world. The Warm Temperate Biome (0.6 ha) provides an experience of Mediterranean climates, including the Mediterranean Basin, California, and South Africa, as well as another Cornucopia area. The Outdoor Biome (12 ha) takes advantage of the mild Cornish winters, which allow for the survival of some very sensitive plants, including tea, and hardier Mediterranean plants and succulents, as well as bananas. These more intensively managed exhibit areas are set off by structural planting schemes that will take many years to achieve their full potential. The periphery is populated with semi-natural plantings, among which is Wild Cornwall, a pastiche of local land-use features, illustrating, from a local perspective, that sustainability, as the saying goes, begins at home. Hydroseeded native vegetation, including gorse, has now become established and blends the highly cultivated Eden site with the surrounding Cornish landscape.

The Core is the most recent addition to the site: completed in 2005, it is the educational-resource building and home of Eden's educational mission. In addition to housing public exhibits relating to how plants influence the world, it contains flexible meeting and teaching rooms and offices. The building itself is inspired by nature, in that its design mimics one of nature's fundamental forms: the intersecting spirals found at the center of, for example, sunflowers or pine cones. The Core's construction incorporated sustainable practices in design, operation, and materials use. Rio Tinto provided the copper roof with metal from its Bingham Canyon mine in Utah, a site widely recognized for its high environmental and social standards. The complete supply trail – from production and processing to transport and fabrication – was documented, demonstrating that it is feasible to account for origins when sourcing copper in this way. This "rock to roof" story has since provided the umbrella for a research project that is investigating the prospects for a system, parallel to that in the timber industry, of certifying metals in construction (see Figures 9.3 and 9.4).

Proposals are currently being developed for the funding of the final phase of Eden's building infrastructure – The Edge – which will have two focal points. The first will be the theme of "transformations," showing how humankind's ancestors adapted their skills in response to extreme climate and ecological change. The second theme will be "Living Within Limits," addressing the arid and semi-arid regions of the world, where climate-, energy-, and water-related tensions are experienced in the most extreme forms and where the corresponding

Figure 9.3 The Core – Eden's education building.

Figure 9.4 The Eden Project 2006.

extremes of wealth and poverty, plenitude and scarcity, live side by side. It has always been important for Eden to integrate with the existing tourism infrastructure, including accommodation providers, as well as the retail, catering, and transport industries, in order to maximize the project's regeneration footprint.

It was apparent from the start that delivering a project of the scale and ambition of Eden was going to be an expensive venture. The total cost, since 1998, has now reached £133 million, paid for through the financial backing of the National Lottery/Millennium Commission (£56 million), the public sector (£25 million from the European Union (EU); £21 million from the United Kingdom), commercial loans (£19 million), and self-financing (£12 million). It would not have been possible to construct Eden, nor to maintain its educational and regeneration ethos, without the significant support of public funds.

The transformational benefits that Eden has brought to Cornwall have dramatically exceeded the original expectations. It has become an icon for the repositioning of Cornwall and the southwest of England, in national and international perceptions, as a modern economy. This repositioning has maintained, if not strengthened, the buy-in from the local community and enabled Eden to develop even more ambitious plans for the site. The project has developed as a major force in the county, not simply in economic terms, but also in terms of the influence of its enterprising approach to waste-management issues, transport, education, horticulture, and accessibility issues. Eden has been a phenomenal success beyond all expectations. In its first five years, the "Eden effect" has:

- opened a spectacular and internationally acclaimed visitor destination and educational resource – on time and within budget;
- injected more than £750 million of value-added (that is, excluding direct spending at Eden) into the local economy, each year paying back in real terms the total of the public's funding investment;

- attracted more than eight million visitors;
- directly created five hundred jobs, with particular emphasis on career development for locals;
- hosted more than a hundred thousand schoolchildren for bespoke educational programs;
- helped to change the perception of Cornwall as a quaint place with pleasant beaches, declining primary industries, and a growing retired population to the perception that Cornwall is "the happening place in the United Kingdom" (population decline among sixteen- to thirty-five-year-olds is gradually reversing, as locals are staying and new residents have migrated to the area);
- supported two thousand five hundred local suppliers (e.g. 83 percent of its catering supplies come from within the county); and
- become a symbol to the environment movement of a can-do attitude.

Eden also strives to maximize its beneficial impacts by "walking the talk" when it comes to sustainable-development issues, making considered solutions, rather than quick fixes, that are viable in the long term and communicable to others. Eden provides an excellent example of the re-use of a Brownfields site for a major transformational regeneration project. A china-clay pit was chosen to show that people can have a positive impact on their surroundings and as an antidote to the constant and consistent bombardment of environmental doom and gloom purveyed daily by the mass media. The regeneration lessons learned from Eden are applicable to anyone or any community or organization, as well as to the mining sector, and ought to inspire these various groups' action in their respective spheres of influence.

A partnership approach was developed initially during the main construction period and has continued into the operational phase. The Eden Foundation – the part of Eden that underpins its charitable and educational aspirations – develops relationships with external organizations and is focused on delivering solutions on the ground through common agendas. Eden's partnership with multinational mining company Rio Tinto grew from modest, uncertain beginnings. Focusing initially on building trust and developing a common agenda around the intellectual space where mining, environment, and society meet, the partnership established a program of activities with the following objectives:

- to enhance understanding of the minerals industries' role in society, earth processes, and sustainable development; and
- to develop initiatives to drive better performance by the mining industry.

Post-Mining Alliance

The Post-Mining Alliance was conceived and remains supported by the Eden–Rio Tinto partnership. The mission of the Post-Mining Alliance is to encourage and promote the regeneration of old mine sites for the sustainable benefit of their respective communities and natural environments. One of its key activities is showcasing success stories and models of good practice from around the world to a broad audience while promoting cooperation and learning from one site to another. The Post-Mining Alliance represents a unique opportunity to bring together mining and regeneration specialists in the interest of community restructuring and regeneration.

With careful planning, a mining operation has enormous potential to contribute to the sustainable development of a region. The real challenge comes when the mine closes and the local community is faced with potential large environmental liabilities and possible socioeconomic collapse. Until recently, mine-decommissioning and -closure activities were not obligatory in most countries. Centuries of inadequate and nonexistent mine-closure practice have left a huge legacy of derelict sites and impoverished communities. This legacy affects the reputation of the entire mining industry, arguably hampering constructive debate and progress in other aspects of its environmental and social performance.

In terms of human impact, the problems of mining legacy are worldwide. Over the last three decades, the virtual death of the coal-mining industry in the United Kingdom has resulted in the immediate loss of more than two hundred and fifty thousand miners' jobs, while millions more have been indirectly affected, either through family or localized business connections. At the other end of the globe, in South Africa, three hundred and forty thousand gold miners were laid off between 1989 and 2004, and the number of people indirectly affected by these lay-offs is significantly larger than that in the United Kingdom.

The Post-Mining Alliance aims to become a repository of knowledge about successful post-mining regeneration projects. It is building a global coalition of interested organizations – in the industry, the government, the non-governmental political sector, in communities, and among Brownfields experts – who are looking for ways to address post-mining regeneration more constructively. Building on the experience of the Eden Project, the Alliance wants to demonstrate that fresh approaches and successful partnerships can deliver positive outcomes in addressing the mining legacy.

The underlying philosophy of the Alliance is that mine closure and the adverse impacts of mining legacy should be addressed by a wider constituency than that of the mining companies alone. Significant regeneration activity is being undertaken worldwide, yet there is still a clear need for coordination in identifying, collating, and disseminating good practice, as well as for coordination in integrating social, economic, and environmental factors.

The Alliance aims to catalyze action on mining legacies, such as converting liabilities to opportunities, stimulating the uptake of good practice in integrated mine closure and mine-closure planning, and encouraging solutions where risks, responsibilities, and opportunities are shared by a wider constituency than mining companies alone. This work involves both documenting and communicating what others have done in an accessible way and looking for opportunities to facilitate discussions that allow diverse local groups to understand how they can work better together to effect regeneration. It has a regional focus, building on existing initiatives, and is based on a mix of policy-level advocacy and project-level activity. The Alliance operates in different arenas – from participating in international public policy dialogues to working on specific projects at a site level.

From the work completed to date, it has emerged that better ways are needed to coordinate existing initiatives and to transfer information globally, that a brokering role needs to be developed, which builds on existing networks while providing leadership for more effective global engagement. Many people remain wary of raising the topic of mining legacy because the issues often seem intractable; the liability issue, for example, looms large, particularly where it is considered the responsibility of the industry rather than of all parties. The Alliance provides a valuable service if it can develop a mechanism that allows this discussion to advance.

There are all sorts of factors that influence successful closure outcomes at a site level, and they vary from region to region:

- a community's priorities and trade-offs;
- the structure of the local economy;
- the extent of interdependence and integration of the mine with the region;
- the local population's capacities and residual skills base;
- the availability of funds; and
- the extent of public-sector and legislative support.

Lessons from the casework collected to date suggest that success in dealing with mining legacy draws on local knowledge, strengths, culture, and capacities. Individual leadership, vision, and commitment are essential, while unusual and unlikely partnerships can deliver excellent results. Mine closure is more likely to be successful when closure is an integral part of the business planning from the design phase, when rehabilitation is carried out progressively, and when partnerships with the local community and post-closure responsibilities are established during operation. Financial mechanisms must allow for sufficient resources, post-closure development initiatives, transferable skills training, and provision for long-term management and monitoring. These success factors underlie all of the Alliance's activities, which can best be described according to five program areas:

1. promoting good practice by developing demonstration models and compiling exemplary case studies from around the world;
2. facilitating and convening events that develop successful approaches to post-mining regeneration;
3. networking with regional centers of excellence;
4. working with partners to advance site-specific solutions;
5. educating and engaging the public through activities at the Eden Project.

Promoting good practice by developing demonstration models and compiling exemplary case studies from around the world

Wider adoption of good practice in mine closure and dealing with mining legacy requires an understanding of what is constraining the adoption of good practice. In general terms, an integrated approach requires a multidisciplinary team that is involved in the mine-planning decisions and risk assessments from the beginning. Some closure standards have been developed, but industry, regulators, and others are increasingly calling for more guidance and capacity building. The methodologies, tools, skills, and success factors required to ensure good practice in closure and new approaches to mining legacy need to be more easily identified, understood, acknowledged, and implemented.

There is some casework to build on, but its documentation and presentation is largely inaccessible. The Alliance now has some information on more than fifty sites, displaying varying degrees of success in post-mining regeneration. Both mines that have been long closed (often abandoned) and sites that are currently undergoing closure have been investigated. In each case, the emphasis is on the lessons that can be captured, transferred, and implemented elsewhere. As this data collection continues, more case studies will be added to the Web-based database.

Most people that the Alliance has approached have responded favorably to the project, and its emphasis on the socioeconomic aspect of closure practice seems to resonate well with their experience; there are already a number of international initiatives focused on long-term environmental legacies. At the same time, it is becoming clear that there are differing perspectives on what is working with respect to post-mining regeneration and, in several cases, there is lack of consensus about what constitutes success. It is expected that these perspectives will be explored in more detail through workshops and further contact with the proponents.

Facilitating and convening events that develop successful approaches to post-mining regeneration

The Eden Project is one remarkable example of post-mining regeneration, and it is hoped that it can be used as a platform from which to inspire broader action on mining regeneration. Using Eden as the backdrop, the Alliance brings together groups who may not have recognized their shared interests nor worked in partnership before.

One strand of work looks at hosting workshops on viable post-mining options. The first module has focused on using tourism as a springboard for post-mining regeneration. The Alliance hosted a weeklong workshop in Cornwall in June 2006 to explore the opportunities and challenges for tourism – both heritage-related and of broader leisure interest – in post-mining areas. This event showcased to a European audience Cornwall's and, more broadly, the United Kingdom's post-mining experience and expertise. This outreach work is being extended to other key mining regions in Australia, Southern Africa, and the Americas. Many imagine that tourism offers a post-mining panacea, and this work will highlight what has worked and what has not and the lessons that have been learned.

A second example of work in this area is a plan for a convening role on the specific component relating to mining legacy in the dialogue on biodiversity between the International Council on Mining and Metals (ICMM) and the World Conservation Union (IUCN).

Networking with regional centers of excellence

There are a variety of organizations, government agencies, industry groups, research institutes, consultants, and others working on post-mining regeneration worldwide. The Alliance wants to engage effectively with these groups and examine where its experience and perspective will add value. A couple of examples highlight its potential in this area.

The Alliance is an active participant in an EU-funded project to establish an effective network of regional public bodies with interests in the mining industry: the European Network of Mining Regions. The aim of the network, which has twenty partners in ten countries, is to develop a common stance on mining and regional sustainable development by using an agreed-upon roadmap and to provide a more coherent multi-stakeholder voice for the mining sector in EU policy circles. There is particular interest among the existing network partners on post-mining issues.

In South Africa, for example, the Alliance co-hosted a workshop on mine closure and legacy with the Centre for Sustainability in Mining and Industry, University of Witwatersrand. The plan is to build on the success of this kind of regional cooperation by working with regional centers of excellence to bring an international dimension to their activities.

Other important regional organizations include the National Orphaned and Abandoned Mines Initiative in Canada; the U.S. Office of Surface Mining; the Center of the American West at the University of Colorado; the Mackay School of Mines at the University of Nevada; the Australian Centre for Minerals Extension and Research; the Sustainable Minerals Institute at the University of Queensland; the University of Western Australia; the Chilean Copper Commission; as well as smaller grassroots and local groups. In addition to institutional expertise, a number of individuals with specific expertise in different regions are exploring potential projects with the Alliance at a regional level. The organization seeks to undertake this regional work with local partners on a project-by-project basis and in the long term intends to solidify these partnership arrangements.

Working with partners to advance site-specific solutions

A key part of post-mining success depends on the ability of the stakeholders to focus on developing and promoting grassroots partnerships and projects that harness creativity and innovation to address the problem of mining legacy. The Alliance is working with local partners to oversee, manage, and deliver specific projects based on the lessons it has learned from observing good practice at other successful sites.

A large challenge presents itself in Cornwall, right on the doorstep of the Eden Project. Once renowned for its mining and engineering industries, the towns of Camborne, Pool, and Redruth now suffer from high unemployment and a culture of dependency, coupled with further retrenchment in the china-clay industry nearby. To date, regeneration attempts have had very limited success. The Alliance is beginning to work with a number of Eden Project partners, drawing on their expertise and experience to engage the community in a series of activities that honor its role in cultural events, its social assets, and its related narratives as a necessary accompaniment to building infrastructure and attracting new business. It is hoped that these partnerships and the activities they develop will demonstrate the benefits of using local communities' often untapped resources to deliver successful regeneration.

Educating and engaging the public through activities at the Eden Project

The Alliance believes that education and better communication of the issues is an essential tool for raising the level of performance in post-mining regeneration activities. Success is often limited by a lack of capacity and know-how – particularly in the socioeconomic aspects of reclamation – among industry companies and the relevant authorities and civil groups. Part of the Alliance's program is the engagement of some of the key mining educators in the creation of flexible learning opportunities for those interested in regeneration. Experiential learning will be a key aspect of this approach. On site at Eden there is already an opportunity to design exhibits that highlight the story of this reclaimed china-clay pit in the context of the broader issues of post-mining regeneration. In addition, there is underway an Eden Project-style publication with the working title *101 Things to Do with a Hole in the Ground*.

Conclusion

The Eden Project provides an excellent platform from which to talk about post-mining regeneration in a constructive and forward-looking way. The establishment of the Post-Mining Alliance provides the vehicle to bring together mining and regeneration specialists to push the frontier on accepted practice.

Who should be involved? Governments should get involved, because the Alliance provides access to a unique worldwide network of practical expertise and experience in post-mining regeneration. Mining companies should get involved, because the Alliance promotes the application, through partnership, of the industry's best knowledge about rehabilitation and regeneration to old and legacy mine sites. Society has a general mistrust of the mining industry and this is often given voice through pointing at the negative legacy left by the industry at so many sites through the centuries. Non-governmental organizations should get involved, because the Alliance can facilitate partnerships that bring the best technical expertise and government support to bear on local legacy cases. Finally, intergovernmental organizations should get involved, because the Alliance is committed to brokering effective relationships among the global mining industries, governments, and civil societies in order to address a worldwide problem.

Chapter 10
Community-based reclamation of abandoned mine lands in the Animas River watershed, San Juan County, Colorado

William Simon

Any successful mine-remediation project is normally dependent upon a thorough characterization program, followed by an analysis of the implementation feasibility of a select assortment of best management practices. Projects on abandoned mine lands often involve a large array of environmental, ownership, responsibility, liability, regulatory-agency, and political-subdivision issues that greatly complicate the characterization, feasibility, and implementation of remediation. Collaborative stakeholder processes, such as that of the Animas River Stakeholders Group (ARSG), which uses a "watershed approach," have proven to be successful in addressing this complexity of issues.[1]

Place, history, and community

The Animas River watershed occupies six hundred and ninety-two square miles of the southwest corner of Colorado, close to the "Four Corners," where Colorado, Utah, Arizona, and New Mexico meet. The Upper Animas watershed is in the heart of the world's largest volcanic mountain range, which includes numerous calderas that are highly mineralized and have been heavily mined for metals. The combination of high elevation (three peaks reach over fourteen thousand feet above sea level), relatively high precipitation, low buffering capacity, extensive mineralization, and historic mining has resulted in many streams where the metal and acid toxicity supports little to no aquatic life.

The area is remote. The Upper Animas watershed, with a population of five hundred and seventy-seven citizens, residing mostly in the town of Silverton, located within the Silverton Caldera, is home to one of the most beautiful, if one of the most difficult, living environments. Since 86 percent of San Juan County is federal land, the remainder being mostly patented mine claims, the tax base is very small. After one hundred and twenty years of hard-rock mining, the last mine closed in 1991, leaving behind unique historic features, a mining heritage, high unemployment, and a legacy of pollution. More than three thousand patented claims contain at least fifteen hundred mining disturbances with poor or no vehicular access and lacking the utilities necessary for "active" treatment systems. Because the Animas River transects Colorado's immense Weminuche Wilderness area, inaccessibility, high elevations, and avalanche dangers make monitoring and characterization of the watershed quite a challenge.

Motivation and challenges

In 1994, the State of Colorado, assisted by the Environmental Protection Agency (EPA), determined that something had to be done to bring the Animas River watershed into compliance with the "fishable and swimmable" requirements of the Clean Water Act (CWA). Previously, many stream segments had slipped through the standard-setting process in being designated as having "ambient quality" with no aquatic-life use. After some initial storm-water controls were implemented at mine sites in the late 1980s, however, followed by brook-trout introductions administered by a volunteer group, the Colorado Water Quality Control Division (WQCD) became aware that improvements were possible and that aquatic-life classifications were needed to protect what minimal biota still remained. The WQCD visited the community to propose the adoption of new classifications and goal-based standards. Many locals, supported by evidence from mining companies, considered the proposals to be impractical and unattainable, reasoning that most streams were toxic primarily as a consequence of natural geological processes. Historically designated names such as Mineral and Cement Creeks reinforced these notions. Many believed that attempts to improve water quality would be futile. An EPA threat to list the entire watershed as a Superfund site, however, resulted in the coalescence of all the stakeholders into a collaborative effort. Scientific investigations and the "watershed approach" would determine the sources, transport, and fate of metal pollutants, as well as the practicality of improving stream water quality and aquatic life. There was a feeling that the Superfund designation would have adverse economic consequences on the county, which was barely surviving while developing a tourism-based commerce. Equally important was the community's desire to avoid the often secretive and impractical legal environment for which the Comprehensive Environmental Response, Compensation, and Liability Act (CERCLA) had become known among Colorado mining towns. Residents wanted to be intimately involved with the science and the decisions that would be made. Thus, the ARSG was initiated.

In a 1995 ruling, the Colorado Water Quality Control Commission (WQCC) chose to delay the implementation of goal-based standards, instead challenging the newly formed ARSG to complete a comprehensive characterization of aquatic conditions throughout the watershed and to make recommendations for the crafting of practical and attainable stream standards and use classifications. The three-year period allotted for this endeavor, later extended by another three years, concluded with a rule-making hearing that relied on the results of the group's comprehensive "Use Attainability Analysis for the Animas River Watershed" (UAA), a synthesis of six years' data collection and analysis, remediation-feasibility analysis, and recommendations for standards that were substantiated by a combination of stream biological potentials and remediation goals (Simon et al. 2001). Temporary modifications to the adopted standards, renewable if needed every five years for up to twenty years, are designed to provide time for remediation and gradual improvements. The document resulted from thousands of hours of volunteer monitoring, investigation, experimentation, analysis, and discussion. At the time of this publication, we estimate that the volunteers' in-kind contributions to the entire collaborative process has been in excess of three million dollars. This dedication has led to a sense of ownership in the process and a stewardship for our watershed resources.

Upon accepting the Commission's challenge, the ARSG immediately retained William Simon as coordinator and restoration ecologist, compiled a watershed bibliography, and developed a plan of attack. Coincidentally, the U.S. Congress had just initiated its Abandoned Mine Lands (AML) program, which designated the Animas watershed as one of two national pilot sites. The program funded the United States Geological Survey (USGS) to do much-needed research on many aspects of acid-rock drainage that would not have been possible or practical for a stakeholders group to fund or accomplish. Although the research contributed minimally to the UAA, it fulfilled the need for a broader understanding of hard-rock mining's impact and to develop characterization methods that could be applied to other watersheds. USGS, Bureau of Land Management (BLM), and United States Forest Service (USFS) – already active participants in the stakeholder process – all benefited from the congressionally funded AML program.

Organizational structure and process

The ARSG chose to remain an *ad hoc* entity, without bylaws or voting privileges, so that everyone in the community, including government agencies, mining corporations, advocacy groups, and landowners, would have an equal voice. The mission was simply "to improve water quality and aquatic habitat throughout the Animas Watershed." The primary goal was to meet the Commission's challenge, but important objectives included natural, cultural, and historical preservation. All stakeholder activities have developed through consensus building, self-empowerment, and a commitment to using scientific methodology. It became obvious that much more information was needed before informed decisions could be made about the adoption of appropriate stream standards, remediation targets, and metal-reduction goals.

Source, transport, and fate characterization

In 1995, a program was initiated to characterize the physical, chemical, and biological condition of all streams and mine sites throughout the watershed. Metal and acid sources needed to be identified and quantified; then a feasibility analysis for remediation needed to be completed for each mine-related source. Because there was much contention over natural-loading versus mine-related sources of contamination, methods were developed to provide reasonable estimates for the contributions of both.

Synoptic sampling within each sub-basin was completed by several dozen volunteers during every water-related event. Eventually, each draining mine's contribution was quantified, and each mine-waste pile was sampled and leach-tested, to estimate its metal-loading potential. One hundred and seventy-four draining mines were ranked with respect to one another for the contributions of each metal and acidity. One hundred and eighty-five mine-waste sites were ranked in a similar fashion. The results indicated that more than 90 percent of all mine-related metal loading was derived from thirty-four draining adits and thirty-three mine-waste sites (Simon *et al.* 2001: XI: 1–7). Combining the sites of lower rank with the sites not fully characterized but evaluated based upon size, location, mineral content, and the lack thereof, the ARSG eliminated more than fourteen hundred sites from imminent cleanup consideration.

During the stable low-flow period of December through February, when surface contributions to stream flow are negligible, the stream sources were attributed primarily to groundwater and mine discharges (Simon *et al.* 2001: VII: 7–9). Since mine discharges had been quantified, the remaining metal loads were considered to derive from natural weathering processes. Although there may have been other possible sources, the stakeholders were satisfied that they were insignificant and impractical to quantify. The results indicated that, while iron and aluminum loads were predominantly related to natural weathering processes, cadmium, copper, manganese, lead, and zinc loads were related to the impacts from extensively mined narrow, enriched ore veins. This conclusion became valuable information, which was then used to set attainable stream standards. It was apparent that a few stream reaches could not be expected to support aquatic biota, given the natural toxic condition created by aluminum and iron. On the other hand, remediation of other metal sources could have beneficial impacts on most streams, especially if the aluminum and iron were diluted or attenuated below the headwaters. Twelve stream tracer studies, metal-speciation studies, and detailed geological mapping completed by the BLM, USGS, and Colorado Division of Minerals and Geology (DMG) provided further evidence for these conclusions (Kimball *et al.* 2002: 1183–207).

Feasibility to remediate sources

Coinciding with the source characterization studies, the physical, biological, and environmental data were collected for each site. This information was to be used in determining remediation possibilities, applicable best management practices, potential for metal-loading reductions, and a first estimation of cost (Simon *et al.* 2001: X: Appendix A–D, 1–5; XI: 4–5). An individual feasibility report for each sub-basin was developed, and later spreadsheet matrices were used to combine the mine-site ranking data with the feasibility parameters, including alternative scenarios for remediation (Simon 2001: Appendix 11A). ARSG chose a fluid prioritization process, whereby the sixty-seven sites would be evaluated periodically to assess the current status of such factors as funding possibilities, owner cooperation, and technology availability.

Determinations of biological potential

Although other watersheds, particularly within the San Juan Mountain range, are affected by mined volcanic calderas, the Animas watershed has been mined the most extensively and its streams the most significantly impacted. Unfortunately, there does not seem to be a highly mineralized, unaffected caldera, which could serve as an appropriate reference for the biological potentials for high natural metal concentrations at the same lofty elevation. For this reason, water from various Animas stream segments was used to conduct biotoxicity studies on likely fish and several macroinvertebrate species, in order to determine these species' threshold tolerances for the dissolved metals of concern (Simon et al. 2001: VI: 11–21).

Bioassays of fish and macroinvertebrates below Silverton, where aquatic life first becomes established, indicated the bioaccumulation of some metals but no human consumptive health risk (Besser and Leib 1999). Although predator–prey relationships were not investigated, it is likely that bioaccumulation may impact the overall diversity of macroinvertebrates.

In the fall of 1996 and the spring and fall of 1997, an extensive inventory of macroinvertebrates and the physical conditions of streambeds was completed to establish the pre-remediation condition. Sites were selected above and below major tributaries of the three sub-basin streams. The tributaries were also sampled at their mouth; a few Animas tributaries, flowing from outside the caldera, could then be used as unaffected reference sites for comparison purposes. The macroinvertebrate inventory was repeated in the fall of 2005, enabling a pre- and post-remediation comparison (Anderson 2006).

Use Attainability Analysis

The 2001 UAA synthesized all of the relevant scientific investigations with the ultimate goal of determining appropriate and attainable stream standards for the entire basin. In addition, baseline conditions and sampling protocols were described and established for the geochemical, fish, macroinvertebrate, and physical parameters of each stream segment. By considering both the biological potentials of the receiving streams and the potentials for metals reduction through partial remediation of the highest-ranking mine sites, standards and aquatic-use classifications were recommended for adoption by the WQCC (Simon et al. 2001: XII: 1–15). These recommendations were approved by the Commission and by the EPA. The standards, depicted as monthly concentrations, were later converted to loads, using ARSG flow-modeling data, such that twenty-nine total maximum daily loads were determined. The Commission approved temporary modifications to the new standards, to be reviewed and possibly modified every five years. The CWA allows up to twenty years for goal attainment.

Remediation implementation

The ARSG did not begin site remediation seriously until after all the mine sites were ranked for their metal contributions. However, when the Sunnyside Gold Corporation (SGC) ceased mining operations in 1991, it immediately initiated its reclamation obligations and eliminating discharges from its mine portals in the attempt to avoid perpetual mine-drainage treatment. In response to a dispute with regulatory authorities over concerns that the springs and seeps resulting from mine-bulkhead placement might require new discharge permits, a court-ordered consent decree was imposed. The decree stipulated that, following bulkhead installation and a five-year period of stable hydrological conditions in the flooded mine workings, there was to be no net increase in metal concentrations at a given Animas River reference location below Silverton. Zinc was used as an indicator element, because it tends to remain dissolved the longest of all the metals of concern. If this condition were met, new or increased metal loads at springs and seeps would not require discharge permits. SGC therefore went about installing numerous bulkheads; but they also remediated several mine sites for which they had no previous responsibility, in hopes that the reduction of contaminants from those sources would more easily assure the required condition at the reference location. Whereas the company could, alternatively, have filed for bankruptcy, they instead undertook the remediation of several sites that ARSG would have found difficult to fund. After spending more than $20 million on the remediation of both their own sites and those of other parties, SGC met the terms of the consent decree in 2002. In so doing, SGC remediated some of the most highly contaminated abandoned-mine sites, which helped meet the stipulation of "no net increase in metal concentration." Unfortunately, since the termination of the decree in 2002, increases in zinc concentrations have been recorded. ARSG is presently investigating whether the increases are the result of a response lag time or the result of other environmental variables.

In 1999, after completing the ranking of the mine sites, ARSG began to remediate mine-waste sites. Two years prior to this endeavor, the stakeholders applied for and received funds through the Section 319 Nonpoint Source program. They discovered that funding cycles, bureaucratic hang-ups, and the extremely short construction season typically add up to a minimum of two years' delay, from application to construction. During this period, further engineering, owner cooperation, bid specifications, and competitive bidding can proceed. ARSG generally subcontracts with the DMG, in order to ensure a competitive bidding process and additional site management during construction. Thus far, project costs have ranged from $60,000 to $800,000 per project, but the more expensive "draining mines" have not yet been addressed, on account of liability considerations.

For mine-waste sites, contractor and third party (e.g. ARSG) liability has been minimized through the use of Removal Action Memoranda, using Colorado's CERCLA mandate, entrusted by the EPA. Following project completion, the landowner is still responsible for any remaining environmental issues, while, at least in theory, the remediating third party is not. Draining mines, on the other hand, have serious liability issues, which inhibit third parties from treating the effluent, even if incremental improvements to water quality can be made. The EPA considers draining mines to be "point sources," which, when the rules are enforced, require National Pollutant Discharge Elimination System (NPDES) discharge permits. The large number and frequent remoteness of abandoned draining mines, particularly in the West, has meant that NPDES discharge permits are seldom pursued by regulatory agencies. However, once a third party touches the water, it can be considered an "operator," in which case it becomes liable for obtaining a discharge permit. An NPDES permit requires, at the minimum, that the operator or owner perpetually treat the effluent to meet aquatic life standards. For third parties and owners without deep pockets, this is a severe disincentive to making any improvements, because as soon as one enters the regulatory/enforcement arena, there is no retreat. Treatment-plant capitalization, operation, and maintenance costs can easily exceed a million dollars per year. At the Penn mine in California, a lawsuit filed against a third party, which was responsible for making only partial improvements to water quality on a draining mine site that affected the San Francisco Bay Area water supply, forced that third party to spend millions of dollars to solve a problem it had not created. The lawsuit resulted in the virtual cessation of third parties from addressing draining mines for fear of long-term reprisals.

Although there have been – and currently are – attempts to pass national legislation that would amend the CWA to provide third parties, or "Good Samaritans," with liability relief, none has been or looks to be successful, because Congress is reluctant to modify the Act. In response to this impasse, ARSG has crafted its own legislation, known as the Animas River Good

Samaritan Mined Land Remediation Pilot Project Act (HR 5071). It was introduced into the House of Representatives in March 2006 by Rep. John Salazar of Colorado.[2] If passed, the legislation would apply to the well-defined and limited problems associated with hard-rock mine drainage in the Animas watershed. The pilot program would then serve as a model of the benefits and challenges associated with such a modification to the CWA, could be expanded to other watersheds, and yet contains a sunset clause, which ensures that the legislation be further evaluated by Congress before enactment of a permanent national provision.

In the meantime, ARSG has successfully reduced two mine discharges by implementing water-infiltration controls. Since the stakeholders have not treated the discharge, they feel confident that they cannot be considered an "operator," which would trigger liability consequences. Unfortunately, few mines have discharges that can actually be reduced in such a practical manner. Another option is installing underground bulkheads, which would impede the flow of water to the surface. A bulkhead can flood the mine works, creating an anaerobic environment, which precludes the survival of the oxygen-dependent bacteria that are responsible for the accelerated acid-rock drainage bio-catalytic process. If the acid-rock drainage process is stopped, acid and metal production are minimized. Stopping this bio-catalytic process, at least theoretically, significantly reduces the release of acid and metals into the water. It can be expected, however, that some acid and metals would be released by the slower and arguably more natural geochemical leaching process. Although ARSG is actively investigating underground mine works for bulkhead applicability, working as a third party, we are uncertain of the liability consequences if we proceed to the implementation phase. Our pilot legislation would provide the liability relief necessary for us to proceed with reducing contamination from draining mines by constructing underground bulkheads and passively or actively treating mine drainage.

Conclusions

In summary, the ARSG process has been successful in gathering and analyzing scientific information to further the understanding of the repercussions of historical mining. It has become an important educational experience for the entire mining, regulatory, environmental-advocacy, and landowner communities. When this information and experience is combined with the basic research conducted by the Animas Abandoned Mine Lands Initiative and the USGS Toxics programs, a wealth of knowledge has developed to assist other communities with similar mining impacts. Numerous remediation projects, supported by stakeholder networks that provide positive solutions and liability reductions to landowners and stewards, have led to measurable water-quality improvements in receiving streams. An extensive basin-wide analysis has recently been compared to the pre-remediation baseline condition in the Animas River watershed. The results indicate that significant water-quality and aquatic-life improvements have been made in some, but not all, streams. There remains much to be learned and many more sites to be remediated. Perhaps most important, people have learned to work together and, in fact, to take the initiative to solve a significant environmental problem. In so doing, the participants are becoming true stewards of their land.

Notes

1. Information about ARSG's "watershed approach" is available on the group's website: www.waterinfo.org/arsg
2. The text of the Congressional bill can be viewed on the Internet at: www.govtrack.us/congress/bill.xpd?bill=h109-5071

Bibliography

Anderson, C. (2006) Internal report, Animas River Stakeholders Group, Silverton, Colorado.

Besser, J. and Leib, K. (1999) 'Modeling frequency of occurrence of toxic concentrations of zinc and copper in the Upper Animas River,' Proceedings of the U.S. Geological Survey Toxics Program Technical Meeting, Charleston, South Carolina.

Kimball, B., Runkel, R., Walton-Day, K., and Bencala, K. (2002) 'Assessment of metal loads in watersheds affected by acid mine drainage by using tracer injection and synoptic sampling: Cement Creek, Colorado, USA,' *Applied Geochemistry* 17: 1183–207.

Simon, W. (2001) 'Upper Animas mine site remediation plan,' presentation abstract, Proceedings of the Colorado Geological Survey Annual Conference.

Simon, W., Butler, P., and Owen, R. (2001) 'Use attainability analysis for the Animas River watershed,' unpublished presentation to Colorado Water Quality Control Commission, Animas River Stakeholders Group.

Chapter 11
Case studies of successful reclamation and sustainable development at Kennecott mining sites

Jon Cherry

Introduction

The demand for minerals and metals in today's society is greater than it has ever been throughout the history of the world. Metal use intensity has declined over time on a per capita basis while the total consumption has increased (National Institute for Public Health and the Environment 1999: 7). This record demand, production, and consumption is driving the world economy and improving the standard of living in most parts of the world. With this increased demand comes increased responsibility to produce those minerals and metals in a way "that meets the needs of the present without compromising the ability of future generations to meet their own needs" (Brundtland Commission 1987: 51). By effectively applying the principles of sustainable development, mining in the twenty-first century can create positive environmental and economic legacies, the impact of which lasts even longer than that of the mines of the last hundred years.

Over the last several thousand years, societies have been defined by the minerals and metals they used: the Stone Age, the Iron Age, the Bronze Age. Even in today's high-tech and mobile society, minerals and metals provide the basic building blocks and infrastructure that make it the healthiest, most efficient, and most productive in history. The minerals and metals that have been a part of human history from the beginning have always come from the earth, just as they do today. The difference is in how the minerals and metals are mined and in the resulting legacies or opportunities created by the mining itself.

By its very nature, mining must disturb the land. In some instances, these large-scale disturbances have resulted in long-term environmental issues, such as acid-mine drainage or subsidence. From a historical perspective, these environmental issues are most likely the result of a lack of understanding about the long-term impacts of mining. In contrast, today's technologies and understanding of science enable present-day mines to be designed with state-of-the-art materials, engineering, and computer modeling before a shovel full of rock is even excavated from the ground. This ability to design and analyze the impact of a particular mining operation before any consequences arise provides mining companies, regulators, and the public with the opportunity to minimize short- and long-term environmental problems, to create lasting social and economic opportunities, as well as compatible land uses for the communities in which the mines are developed.

In 1987, the Brundtland Commission Report defined sustainable development as "that [which] meets the needs of the present without compromising the ability of future generations to meet their own needs" (Brundtland Commission 1987: 51). Sustainable development is the foundation upon which all new mines, their expansions, rehabilitations, and day-to-day projects are evaluated at Kennecott Minerals (a part of Rio Tinto Group) and other mining companies. At a recent conference of Mining and Metallurgical Institutions, Rio Tinto CEO Leigh Clifford challenged the mining industry to "show that modern mining companies are not economic juggernauts but responsible and responsive organizations, conscious of the role they play in society" (May 27, 2002).

The implementation of sustainable development at Kennecott Minerals requires that each project be evaluated according to its four pillars: 1) economic prosperity, 2) environmental protection, 3) social and community well-being, and 4) governance. The first pillar is intuitive, in that economic prosperity requires a project to be financially sound in order to proceed. The second pillar is inseparable from the first, in that all of the economic impacts associated with properly protecting the environment are fully accounted for in a project's economic analysis. In other words, a project is only economically viable when all of the proper environmental-protection measures are in place. This also means that there are some areas where the development of a mine is inappropriate. The third pillar requires that an accounting of the social and community impacts of a project be incorporated into its design. Negative impacts must be minimized, while positive impacts must be identified and exploited to the benefit of the local community, both during the operation of the mine and after it is closed. The fourth and final pillar requires that proper governance be established to review and authorize mining projects, and that the same internal policy and procedures be followed even when local regulations are less stringent than company standards.

It is in this context that Kennecott Minerals and other Rio Tinto Group companies are creating tremendous social and economic opportunities in the design of new mines and at existing mines. They are applying sustainable-development principles to all projects and implementing twenty-first-century materials and engineering technology. To illustrate this process and its outcomes, two case studies are presented below. The first case study is of Kennecott Minerals' Flambeau Mine in Wisconsin, which was designed, constructed, operated, and reclaimed in the 1990s. The second case study is about an active hundred-year-old open-pit copper mine in Salt Lake City, Utah. Kennecott Utah Copper and Kennecott Land Company are building new communities adjacent to the mine operations.

The design is based upon a collaborative approach from overlapping disciplines. In addition to traditional engineers and architects, community planners, transportation planners, biodiversity specialists, energy efficiency specialists (e.g. green building) and landscape architects all contribute to the master design. The key to a successful design is to meld the wants and needs of the community with the various ideas and designs from the design team. Many times this involves sharing new ideas with the community that are different than local customs.

Case study #1: Flambeau Mine, Ladysmith, Wisconsin

After several years of study, design, permitting, and positive interaction with a community mining-oversight board, the Flambeau Mining Company began construction of an open-pit copper mine in 1991. Although the formal sustainable-development program mentioned above had not yet been established at Kennecott at that time, all of the pillars of the now-formalized program were incorporated into what is considered by many to be one of the best examples of sustainable development in mining. After it was determined that the ore body was of sufficient economic value to develop a mine, environmental-impact assessments and mine designs were evaluated for many years before a mine design was agreed upon. The final design minimized the environmental impacts and created a post-mine land use that was compatible with the community's future vision yet maintained the economic viability of the project.

Using the sustainable-development principles discussed earlier, Flambeau created economic wealth not only for Rio Tinto shareholders, but also for the local community. This was accomplished with the implementation of substantial environmental-protection measures, which resulted in a violation-free track record throughout the mine's active operation and in the ten years since its closure and reclamation. In addition to backfilling the open pit and reclaiming the site as an open grassland prairie with mixed wetlands, the design constructed hiking, biking, and equestrian trails for public recreational use. The pre-mining, reclamation concept, active mining, and reclaimed site are shown in Figure 11.1 for a chronological comparison.

With the help of state regulatory agencies and the community mine-oversight board, it was possible to leave in place the Flambeau Mine warehouse, administration building, water-treatment plant buildings, and associated infrastructure. The preservation of the facilities was a cornerstone of long-term economic development for the community; still owned by Kennecott, they are leased to the local economic-development board for a nominal fee. Through the cooperative efforts of the City of Ladysmith, Rusk County, and Kennecott Minerals, $5.7 million in mining-tax revenue was leveraged into an additional $5.7 million in grants for the local community (Northwest Regional Planning Commission 2005: 6–73). In total, the local units of government received more than $11 million, either directly or indirectly, from Kennecott's Flambeau Mining Company. These funds were then reinvested in the community to create economic development opportunities, including 32,550 square feet of new public-building space, 443,450 square feet of new manufacturing or business space, and a new tax base of more than $6.5 million (Northwest Regional Planning Commission 2005: 6–84). This exceptional level of funding to the local units of government leveraged financial support from many public- and private-sector sources, which led to the creation of a new tax base and hundreds of additional jobs. The foresight and planning of the local government, regulatory agencies, and Kennecott preempted at Flambeau the typical environmental legacies and boom-and-bust cycle historically associated with mining projects.

In recognition of these efforts, the United States Department of the Interior, Bureau of Land Management, awarded Kennecott Minerals the inaugural Hardrock Mineral Community Outreach and Economic Security Award in 2004. This award is now presented annually to recognize the efforts made to implement the principles of sustainable development by balancing environmental stewardship, economic prosperity, and social considerations in planning for mining operations.

Figure 11.1 Flambeau Mine chronology.

Case study #2: Daybreak Project, Kennecott Land Company, Salt Lake City, Utah

Although Kennecott Land Company was formed in April 2001, its history in the Salt Lake Valley began with the creation of the Utah Copper Company in 1903. Daniel Jackling founded the large-scale mining company, which was the birth of Utah's rich mining history and the beginning of what is now known as the Bingham Canyon Mine. During World War II, Bingham Canyon produced one-third of all the copper used by the Allies. Kennecott Utah Copper continues to mine and process the rich minerals unearthed in the Bingham Mine and is the second-largest copper producer in the nation today.

During the 1900s, large tracts of land were purchased for potential mining, or as buffer land for the growing community in the Salt Lake Valley. These combined lands now total nearly 93,000 acres along the West Bench of the Valley. Although most of these lands were never mined, some parcels were impacted by adjacent mining activities, which led to a period of unprecedented remediation, restoration, and reclamation in the 1990s. These remediation and restoration efforts cost more than $350 million and included the excavation and removal of 26 million tons of historic-mining waste rock and tailings. Figure 11.2 shows before and after photos of a typical historic-mine reclamation project in Butterfield Canyon adjacent to the mine site.

During this period, Kennecott Utah Copper hired national planning experts to evaluate the potential uses of the reclaimed landscape in the area of South Jordan. In 1999, a team of experts began a detailed evaluation of the potential for land development on the company's entire West Bench land holding. These experts included the firms of Glatting Jackson, Calthorpe Associates, and The Planning Center, which all brought new design and development concepts to Kennecott and the community. Kennecott Utah Copper began working with the City of South Jordan to enact zoning that would allow for a large-scale mixed-use development on 4,126 acres of land. The plan provides for 13,667 residential units, as well as significant commercial entitlements, and more than 1,200 acres of parks and open space for the community to enjoy. These provisions make it the largest master-planned development in Utah's history. The project was named Daybreak and is now home to more than five hundred families. Figure 11.3 shows a site plan of Daybreak.

In 2001, Rio Tinto established Kennecott Development Company to focus exclusively on this new opportunity: developing Kennecott Utah Copper's land and water assets in a way that would create enduring and sustainable communities. The company name was changed to Kennecott Land in November 2002.

Daybreak is just the first step for Kennecott Land, which is planning now for the eventual transformation of the project from a private landholding into a public showcase of new communities and spectacular open spaces over the next 50 to 75 years. The former mining buffer lands on the West Bench of Salt Lake Valley will feature transit centers near homes, jobs for local residents, and schools and neighborhoods with open space connected by miles and miles of trails and parks.

Kennecott Land owns 93,000 acres along the Oquirrh Mountains and foothills, 80,000 of which are on Salt Lake Valley's West Bench. The Valley is projected to grow by 1.2 million people over the next twenty years, primarily as the result of the next generation of Utah families. Kennecott Land believes that the West Bench is the logical place for long-term, quality growth in the Valley, since it owns more than 50 percent of the remaining developable land in the Valley.

To achieve this vision, Kennecott Land is working locally to build enduring communities along the West Bench. This work includes the incorporation of the overarching design principles of sustainable development through the creation of economic prosperity not only for the mining company, but also for the local community, governments, businesses, and contractors that are, or will be, part of the project over the next 50 to 75 years. Environmental stewardship is demonstrated through the reclamation of old mining sites, as well as through the application of quality

Figure 11.2 Butterfield Canyon reclamation, before and after.

growth, community development, and mass transit to the design of new communities. Likewise, social impacts are addressed in the building of enduring communities, such as churches, schools, shopping centers, and businesses. The resulting economic base builds lasting value once mining operations are completed. Governance is addressed through seeking input from local administrators on master plans and by obtaining the proper entitlements, zoning, and permits.

The ultimate vision for Kennecott Land is the development of the West Bench into multiple sustainable communities, as shown in Figures 11.4, 11.5, and 11.6. The West Bench master plan includes the construction of 163,000 residential-housing units; 50 million square feet of new commercial space; mass-transit corridors; miles of recreation trails; and thousands of acres of open space and parks over the next several decades.

In the last century, Kennecott Utah Copper has created thousands of jobs and contributed billions of dollars to the Salt Lake Valley economy. In the next century, its sustainable-development legacy, as developed by Kennecott Land, will be just as significant as – and more enduring than – the Bingham Canyon Mine that started it all.

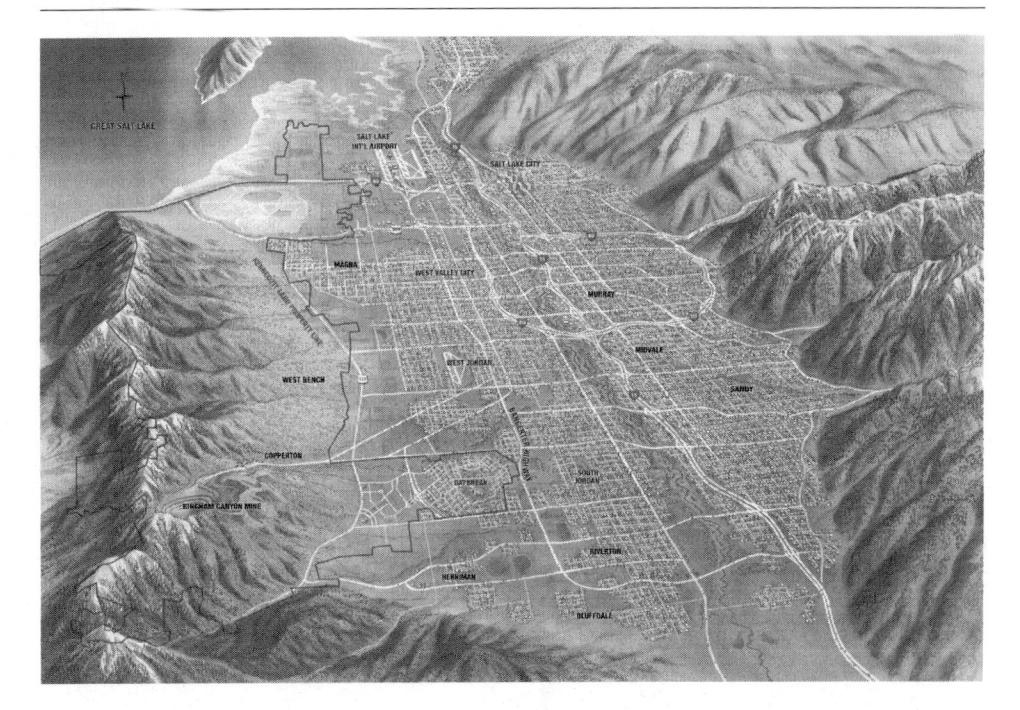

Figure 11.3 Daybreak site plan.

Figure 11.4 West Bench master plan, looking West.

Conclusion

The miners of today have a responsibility to produce minerals and metals that meet the needs of the present without compromising the ability of future generations to meet their own needs. This responsibility is inherently tied to sustainable development. No longer is a mine plan the sole product of geologists and mining engineers that describes how minerals will be recovered from the earth. The mine plan of today is a comprehensive plan that extends beyond the production life of the mine and is the collective production of multi-disciplinary efforts including engineers (civil, mining, environmental), architects, land planners, transportation planners, biodiversity specialists, and landscape architects with significant input from the local community.

Case studies at Kennecott mining sites

Figure 11.5 West Bench master plan, plan view.

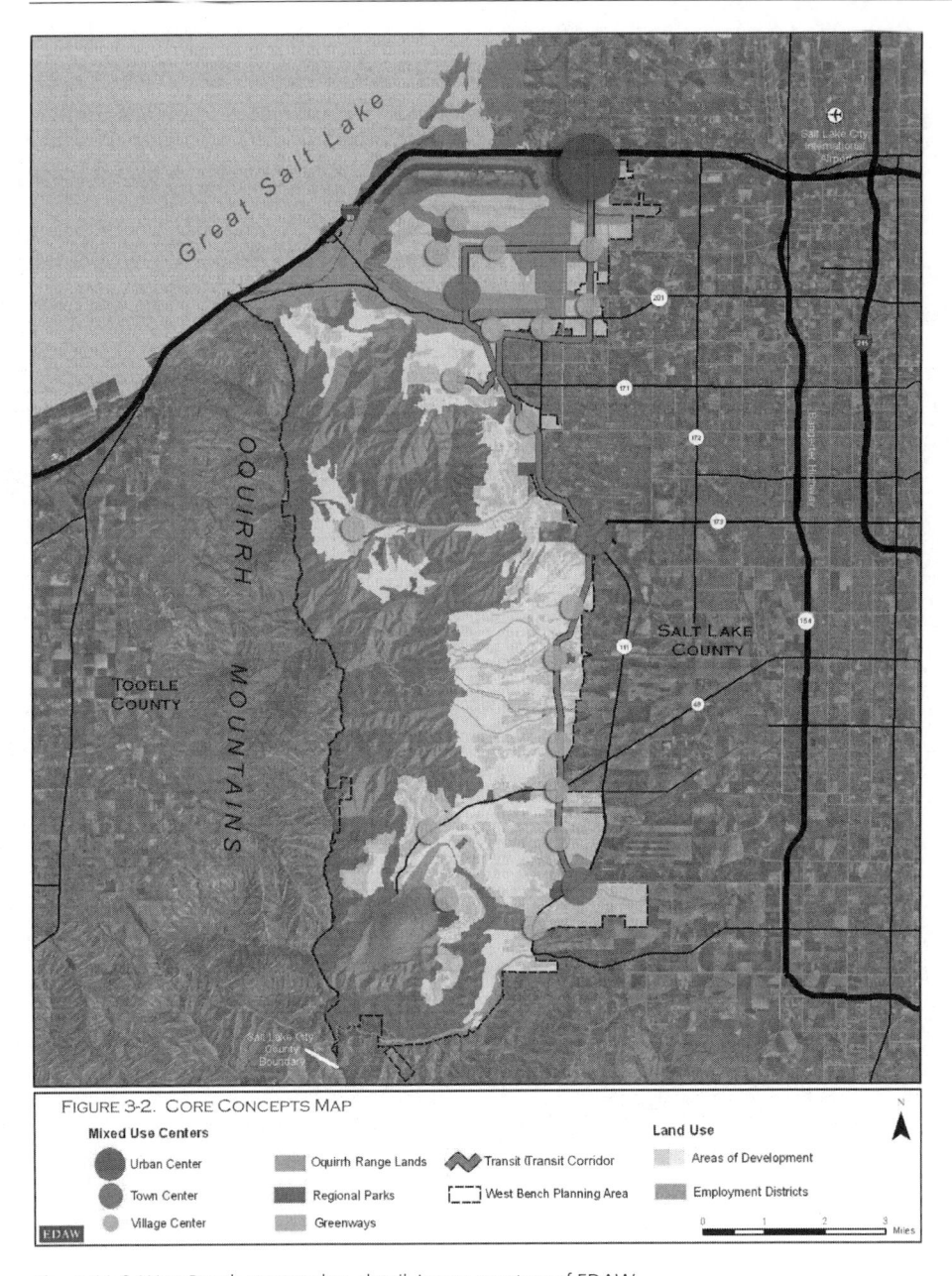

Figure 11.6 West Bench master plan, detail. Image courtesy of EDAW.

Bibliography

Brundtland Commission (1987) *Definition of Sustainable Development*, 51.
National Institute for Public Health and the Environment (1999) *Long-term Perspectives on World Metal Use – A Model Based Approach*, Bilthoven (Netherlands): RIVM, 7.
Northwest Regional Planning Commission (2005) *A Socioeconomic Study of the Flambeau Mine Project*, 6–84.

Part III
Technology, representation, and information in reclamation design

Chapter 12
Digital simulation and reclamation:
strategies for altered landscapes

Alan Berger and Case Brown

Out of sight, out of mind. This was the perception that the mining industry relied on for decades in order to carry out landscape reclamation at minimum expense. Yet, increasingly, people are living closer to active and abandoned mining sites. As mine landscapes and new community developments merge, the mining industry and the public alike will require new modes of representation to communicate their respective interests in the negotiation over the landscape outcomes, or post-mine land uses, of these drastically altered sites. All mines require a post-mine land-use plan as part of the permitting process, even before ground is broken for a new mine. On the one hand, as a result of shifts in environmental values in water, pollution, and noise concerns, and in post-mine land-use desires, the public is becoming more vocal in the permitting process. On the other hand, while the mining industry appears to be interested in addressing concerns of landscape representation, it has, for any number of reasons, yet to develop the dynamic communication tools necessary for dealing with the more complex constituent bases that are now part of the decision-making processes of mine closure and post-mine land use. The average citizen cannot wade through the technical language of Environmental Impact Statements (EISs) and then provide useful comments on the impact of a mining operation and how its post-mine land use is being implemented. More often than not, the post-mine land use and its design are not represented in three-dimensional format, and rarely does the public have an opportunity to critique and adjust the cost–benefit breakdown of reclamation activities. Through new digital-simulation techniques, however, a more substantial public discourse can develop whereby the public feedback and response portions of EISs and environmental assessments can be turned into visual, rather than solely textual, representation.

For the past several years, the Project for Reclamation Excellence (P-REX) has been researching the usefulness of representing and interfacing data in the design of post-mining scenarios. This chapter showcases design research from the French Gulch Simulation Project, funded through a grant from the U.S. Environmental Protection Agency's (EPA's) Superfund Program, through the EPA's Office of Land Revitalization, and through the EPA consultant Tetra Tech, Inc. In what follows, it will be demonstrated how landscape architectural design and representation can more effectively address the ecological and economical challenges of landscape disturbance and renewal. The ultimate goal of this research is to develop a digital environment that leads to greater efficacy and community input in post-extraction or abandoned-mine-land-redevelopment decision making. Although the prototype shown here was created for mined environments, the research and representational strategies can easily be adapted to a variety of landscapes and projects where environmental alterations have occurred. Given the time, resources, and sustained interest in this project, one can just imagine a Web-based real-time interface for post-mine land-use negotiations among communities, regulators, designers, and the mining industry.

Introduction: positioning digital simulation within reclamation design

Designing for landscapes that have been drastically altered by mining processes requires a realignment of traditional landscape-architectural objectives, priorities, and methodologies. In spare economic and ecological environments that have been long since abandoned by bankrupt or dissolved mining companies, normative design practices are of little value. In the simplest sense, altered landscapes are sites where vast quantities of materials have been added or removed from the landscape, resulting in several chemical, biological, and physical changes. Altered landscapes require a more humble recognition of their physical limitations and the fiscal resources needed to enact changes upon them. Designing for altered landscapes therefore requires an adjustment of open-ended design objectives toward optimal landscape performance from a minimal energetic input. Three- and four-dimensional digital simulation can provide a platform from which to explore these objectives and permits a visualization of how the components of a reclamation design may adapt to one another. Reclamation design emphasizes adaptation to, rather than replication of, the environment, thereby suggesting a major distinction from what has been termed "restoration" (Berger 2002: 61, 71).

Software bundling, digital work flow, and interface

When digital tools are used in landscape design, they tend to be employed as substitutes for the older analog operations of drawing. Autodesk Autocad, for example, is now widely used to draw lines for base plans, rather than a T square and pencil. This relationship change merely substitutes one analog mode of representation for another. The proposed digital work flow is radically different. Base data is received or output from Geographic Information Systems (GIS) in digital form, which is not continuous or analog, but rather a series of data points, or vertices. These vertices are then manipulated as three-dimensional fulcrums, which directly determine the three-dimensional model. The advanced digital programs do not understand continuous contours, only vertices, which creates a fundamentally different operation and mode of thinking than those work flows which mimic analog models. It is, no doubt, a substitute for analog, but it is, nevertheless, a digital, discontinuous substitute with different purposes and functions than analog modes.

The standard landscape-design visualization pathway using digital software produces a singular, static final-design image from original-terrain data (see Figure 12.1). These data (i.e. information that yields components of terrain, vegetation, etc.) are essentially made dormant by the lower portion of this pathway. In other words, the vegetation, land surface, and infrastructure are already fixed and no longer capable of dynamic updating, so that this linear process eliminates the innumerable ecological possibilities inherent in a changing, altered landscape. The conditions of such landscapes inevitably change as elements are added or removed, thus rendering a singular, solidified design maladapted.

This new approach attempts to develop a more robust, ecologically based model that utilizes software to generate various sets of conditions, providing the base stratum for evolving landscapes. Currently, reclamation designers (i.e. landscape architects) are the inventors and carriers of those ecological conditions, transferring their knowledge of ecosystem parameters to the physical model. Ideally, one would have a software bundle that understands these conditions and their parameters and could output a series of possibilities based on region-specific eco-data. Short of this ideal, digital work flow, or how data pass from program to program, adds more dynamic function and complexity to the final model (see Figure 12.2). The base contour data can be handled by the industry-standard software: Autocad and the Civil Design Series. ESRI's ArcGIS performs as data-management software for the base contours, parsing out and simpli-

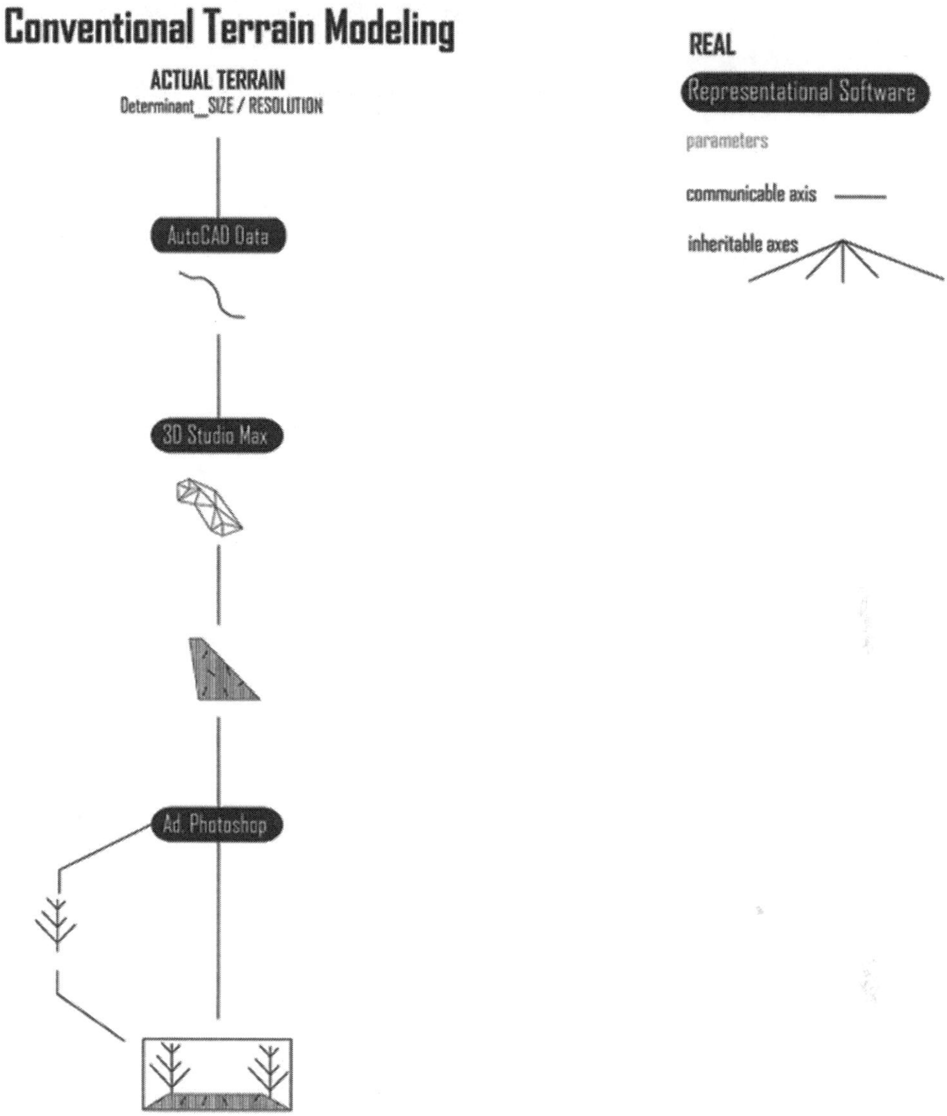

Digital simulation and reclamation

Conventional Terrain Modeling
ACTUAL TERRAIN
Determinant_SIZE / RESOLUTION

AutoCAD Data

3D Studio Max

Ad. Photoshop

REAL

Representational Software

parameters

communicable axis

inheritable axes

Figure 12.1 Conventional terrain modeling flow chart, which is linear and static.

fying them, as well as locating various habitats, edges, and jurisdictions in relation to the site in question. As the landscape-contour stages are modeled, 3D Studio Max performs the function of testing the surfaces and converting them to Triangulated Irregular Networks (TINs), or a collection of nodes and connectors. This efficient surface model can then easily be split into several surfaces indexed to a series of ecological parameters (i.e. south-facing and dry, or elevated metals, or low and perennially wet). As these parameters approximate the operation of the real landscape, the model is imbued with various ecologically determined terrains.

With the landscape surface created and selectively dissected into distinct terrains, the model can be exported to e-on Software's Vue 5 Infinite. This software was conceived to model and

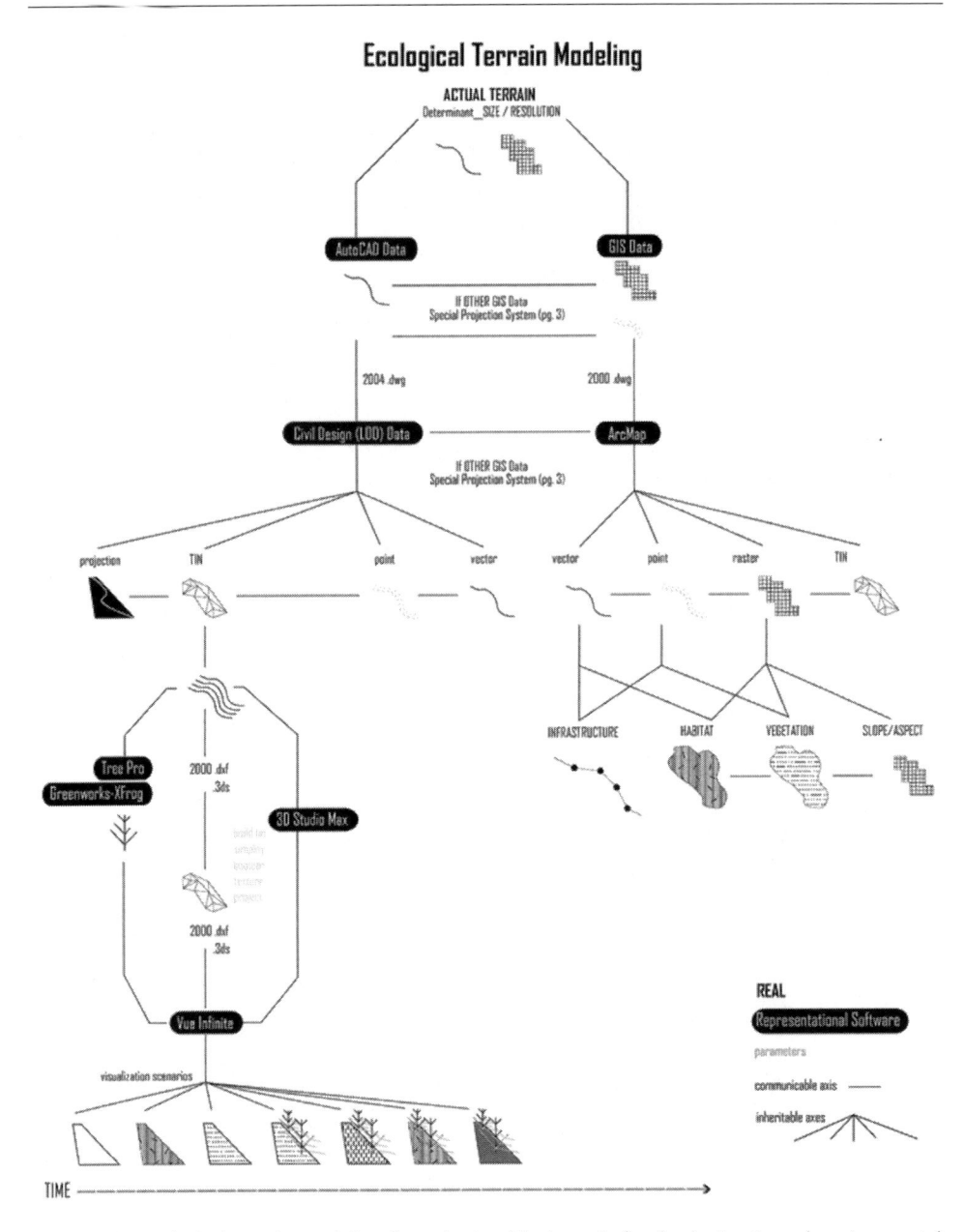

Figure 12.2 Ecological terrain modeling flow chart, with dynamic feedback circuits and environmental system inputs.

animate alien worlds for the movie industry, which wanted more naturalistic computer-generated environments. For the purposes of mine reclamation, Vue can populate the distinctly defined terrains with vegetation according to parameters such as slope, aspect, size, density, and distribution. Its very robust rendering engine and polygon management enables thousands of instances of three-dimensional vegetation to populate the model. Both the rendering engine and the polygon management allow the digital work flow to approximate more effectively the appearance and processes of the natural world. With the input of vegetation modeled in such programs as Tree Pro or Greenworks X-Frog, a hundred billion polygon models are possible in Vue.

From this dynamic work flow, data can be passed in and out of the software as updates become available. Updating the data, however, does not lead to producing new views. The digital work flow is strictly for visually driven design. As an ecologically driven design process, the parameters are changed to produce a different landscape ultimately, or at least a different landscape model in terms of the simulation. The view is updated and unique, of course, but that alone does not justify changing parameters. To focus on views would be to undermine the purpose of this work flow. The proposed methodology expands the designer's capabilities in terms of evolving a design over time and in terms of its physical and structural requirements. The most spectacular advantage to this work flow is that it enables the seeing, designing, and analyzing of the design as a series of morphing stages, rather than as one static, final view or outcome. Multiple future trajectories can be conceived for any given starting point in time. Stakeholders, for instance, can view the landscape in its nascent, developing stages, rather than in its twenty-year maturity. Only in this way can the reclamation designer work with a site's complex ecological systems and physical manipulation while communicating more effectively with the project stakeholders (see Plate 1).

To envision additional nonspatial components of reclamation design, a beta version of a simulation interface has been developed, which includes external project parameters, such as budgeting and phasing (see Plate 2). As a programmable tool, this simulation interface is a dynamic bundling of temporal, material, ecological, and economic variables. Design alternatives can thus materialize from a series of interacting inputs, such as costs, task scheduling, ecological evolution, and human interference. Taken as a whole, the simulation interface, when tied to the dynamic work flow, produces multidimensional landscape models better attuned to the complexities of actual reclamation.

Reclamation-site design principles

Several site-design principles emerge from the constraints of post-mined or altered landscapes. The factors of time management and project cost often far outweigh the historical forces that altered a site's conditions in the first place (Power 1996: 129). Produced in merely a few years or decades an altered landscape may require extensive funds and a century or more before it is reclaimed with a renewed use. Three principles are therefore developed here, in order for readers to grasp the uniqueness of reclamation design and the inherent benefits of approaching this type of practice using the proposed simulation work flow and interface.

The conservation of energy and mass in site operations

The sheer scale and degree of alteration in extractive landscapes negates normative prescriptions for design. Altered landscapes require a more humble recognition of physical and fiscal limits. Therefore, altered landscapes encode a realignment of one's design objectives toward optimal landscape performance from minimal energetic input, or, that is, toward a conservation of energy and mass in site operations.

Several basic axioms emerge in this realignment. First, the beneficial and transformative properties of the actual site materials should be assessed and programmed into the basic design formulations. This assessment may reveal unique or beneficial uses of local materials, enhancing the character of the landscape while reducing transportation and procurement costs. Second, the strictures of the new, altered conditions must be recognized and deterministic of vegetative, programmatic, and structural decisions. Various commonly used landscape plants, for example, may be maladapted to the site's altered aridity, metals load, or periodic inundation. This second axiom can both avoid the costs of replanting and may inspire innovative design configurations for the altered landscape. Third, material movements should approach equilibrium, leaving little extra material and requiring minimal external input. Fill can easily be generated at most sites, while cut can prove more difficult, as it may need to be landfilled, capped, or otherwise consolidated. In the Breckenridge, Colorado, French Gulch Project (see Chapter Eight in this volume) inert dredge rock will be crushed into gravel and resold for construction purposes. Even if on-site materials require crushing and mixing in order to produce the design objectives, they are likely to be cheaper than imported materials. Last, all of the aforementioned considerations interact dynamically with one another, multiplying the functional results of each design objective. Locally diseased timber, for instance, could be chipped into an organic mix to enhance the disturbed soils; or crushed waste rock could be bound into a more inert form to provide a road base. Overall, these basic axioms help turn the material constraints of reclamation planning into generative, synergetic design drivers.

The adaptive use of site conditions

The human-induced and natural processes that shape the surface of an altered landscape offer evidence for designers: what does or does not survive in the post-extraction environment may signify essential ecological gradients for exploitation or modification in the reclamation strategy. Given that site maintenance is likely to be sparse – or even nonexistent in the case of an abandoned mine – the designer must devise approaches that "disturb the disturbance" in order to reintroduce ecological diversity to the site. Instead of replacing a site's problematic soil, for example, a more adaptive approach might include scarifying the surface to catch water and slowly ameliorating the soil with natural processes. In a poor drainage scenario, one might consider designing a new wetland habitat. Soil conditions are usually the prime determinant of the vegetative strategy and planting design. Disturbance-adapted plant species, moreover, have a much greater chance of surviving the early stages of a reclamation project. The designer should conduct a vegetation assessment to determine whether any disturbance-adapted plant species have local populations. Many of these types of plants have pioneering functions, such as nitrogen fixation, or other phytoremedial properties. In scenarios where conditions are unknown, an entirely adaptive approach to planting may prove useful; when inexpensive, seedlings are mass-planted across a site to evaluate which species and varieties can best survive under the altered conditions (Johnson *et al.* 1994: 31, 35).

The biological oddities of post-mined landscapes also offer designers important lessons for site programming and adaptability. The U.S. National Park Service reports that declining bat populations have found new habitats in abandoned mine adits. Nearly two-thirds of all bat species in the USA now roost in abandoned mine shafts (National Park Service 2002). Aquatic and terrestrial microbes, which have major roles in the recycling of biomass and essential elements, also thrive on disturbed sites. Such extreme microorganisms have been found near coal fires, in anaerobic mining excavations, and in highly acidic or alkaline soils and water. The Berkeley pit, in Butte, Montana, which now holds more than 30 billion gallons of toxic water, is home to more than forty microorganisms, which have adapted to the water's high acidity and are flourishing in an environment of limited competition and few predators. Many scientists believe that these "extremophiles" may be able to clean up toxic waste naturally and will then reveal the chemistry for finding cures to our worst diseases. Microbial scientists at Montana State University, for example, have approached the National Science Foundation as well as regulatory authorities about reclaiming the Berkeley-pit site as a national microbial observatory lab, specifically for the study of extremophiles. Re-disturbing an altered site can thus enhance the conditions that enable biological agents to cleanse and prepare the water and soil for recolonization by flora and fauna. Successively, this preparation creates the conditions under which plants and animals thrive, thereby performing ameliorative functions through various nutrient-, water-, and material-recycling processes. In essence, the designer attempts to use the altered site's condition adaptively to guide a series of time-based processes of landscape evolution, rather than imposing a single solution that requires heavy initial investment and perpetual active maintenance.

Interactive landscape circulation and infrastructure

Reclamation always has a long time factor. If the post-mine land use is designed to include activities that allow people to occupy or traverse the site, then the designer should consider how circulation systems, whether pedestrian or vehicular, will be integrated into the reclamation process. People who use a newly reclaimed site are naturally drawn to some areas and repelled by others. The planning and design of new pathways, as well as of accessibility and circulation infrastructures, should be done in such a manner as to reduce the potential for direct engagement with high public-health-risk areas, such as deteriorating structures. These infrastructures may also double in function by creating a narrative experience of the alterations that have taken place. Such a circulation system may link the site's history with its post-mine land use, forging a productive educational experience for users.

Bridges, paths, trails, roads, and other circulation infrastructures can be designed to assist a site's ecological functions. For instance, paths that cut across a site's environmental gradients (i.e. from wet to dry or from toxic to clean) elucidate the anthropomorphic and natural systems whose interaction has shaped the landscape over time. If designed comprehensively, the circulation network becomes an infrastructure that collects, controls, or distributes ecological matter (such as water, wind, animals, or seeds) while providing site access to users. Maximizing the functions of landscape-circulation systems may also justify their inclusion in otherwise restrictive fiscal situations.

Case study: the French Gulch/Wellington Oro reclamation

Application of site-design principles

The French Gulch/Wellington Oro area in Breckenridge, Colorado, was chosen as a test site for the application of P-REX's digital-simulation methodology.[1] French Gulch's unique abandonment and ecological alteration occasioned a reconsideration of digital simulation and the evolutionary roles that can be embedded in representational formats. The area rests in a high alpine setting of the Rocky Mountains, at around ten thousand feet (3,400 meters). Extensive placer mining between 1890 and 1960 dredged up the valley floor, leaving massive piles and wandering lobes of waste rock (gravel) tailings (see Plate 3). These tailings still fill the valley floor, spreading laterally to the toe slope of each side's mountainous incline. Snaking through these mountains are more than fifteen miles of underground tunnels and shafts that currently drain into the valley area and eventually into the various bodies of water downstream. The ability to remediate this water is compromised by some of the chemical and physical properties of the site. Water emerging from underground mine facilities contains acid-mine drainage, which needs to be treated upon daylighting from the earth; and water flowing down the valley corridor, meanwhile, is buried beneath millions of cubic yards of dredge tailings. Both water sources are merely accelerating the transport of pollution toward more populated areas and reservoirs. A stakeholders group of town officials, U.S. EPA managers, and local citizens therefore determined that the project site should lie between a proposed water-treatment facility at the top of the valley (to be built by the U.S. EPA and Breckenridge) and a new housing development at the bottom (see Plate 4).

The stakeholders group prescribed three programmatic goals for the reclamation design: 1) One end of the site in the valley corridor – adjacent to the water-treatment facility at the top of the valley – should function as a gateway to the 2,200 acres of backcountry wilderness and recreational trails used by people throughout the county; 2) The other end of the corridor – adjacent to the new housing development at the bottom of the valley – should function as a landscape amenity and "park-type" setting with passive recreational uses; 3) The middle of the corridor – the bulk of the land falling within the project zone – should be designed to integrate mine-waste cleanup with landscape-architectural design, inclusive of access to historic sites, of water-system and plant-community reclamation, and of universal recreational accessibility to the backcountry open-space network (see Plate 5).

The design strategy that resulted from these objectives also evolved out of several existing physical aspects of the site and out of processes that were already in motion there. The existing dredge rock, which fills the valley, will be used to create topographic diversity. At the time of this publication, some of the dredge rock is being sold, crushed, and hauled away, creating a budgetary surplus for construction and a topographical lowering that will allow the daylighting and remediation of the buried water bodies. The town's historical society prefers this solution because it views the dredge-rock piles as part of the site's historical continuum. Extensive use of the dredge rock also conserves energy and mass in site operations, which, it may be recalled, is the first principle of the P-REX methodology.

After the initial site analysis, it was determined that upstream riparian populations of various sedges (*Carex spp.*) and willows (*Salix spp.*) were healthy and abundant. If, therefore, with careful preparation of the soil and hydrological gradients, the site plan could accept pioneers from this seed bank, a vibrant riparian ecosystem could repopulate with volunteer colonization. The sedge and willow species are naturally adapted to stream environments and would be likely to thrive in other disturbed conditions, as well as to accomplish some of the site reclamation by accumulating sediment and filtering water (United States Department of Agriculture 2002). Further verification of these performative functions is provided by the fertile islands of volunteer vegetation that have emerged along the disturbed stream corridor. Utilizing the upstream populations and their self-organizing propensity more opportunistically adapts the design to the site rather than forcing the site toward some unsustainable state, in keeping with the second P-REX principle.

The post-mine programming of the site is, in fact, being negotiated during the writing of this chapter, and consensus is expected in late 2007. The current stakeholders' plan calls for reprogramming the site with passive open space and interpretive trail systems. While the physical manifestation of these programs will continue to evolve as the negotiations continue, it is clear that the circulation network and interpretive history of the site can be designed as an integrated system, according to the third principle of the P-REX approach. The current design specifies a "light infrastructure" of trails, bridges, ramps, boardwalks, and overlooks that traverse the reclaimed riparian corridor, revealing the mining processes that produced the altered condition, as well as the resurgent ecological processes that are reversing some of the alteration (see Plate 6).

Design-simulation description

P-REX's design strategy employs the site's natural topographic gradients and vegetative associations in order to slow down and filter the water moving through the dredge corridor, from the abandoned mine land down the French Gulch and into the Wellington Oro neighborhood. Using the digital-work-flow methodology in tandem with this design strategy produced simulations at each predictable evolutionary phase of the landscape (see Plate 7). In this way, the realities of reclamation could be better analyzed and debated in the stakeholder meetings. In fact, this strategy did result in the stakeholders' ability to envision and comment on the design and then make adjustments according to their thinking.[2]

The largest operation in this reclamation design is relocating more than 75,000 cubic yards of dredge rock, a crucial initial step toward exhuming the riparian corridor from under the dredge piles and daylighting the water. The stakeholders group identified the expansion of the riparian corridor as a highly valuable component of the design, and the excess material will likely have two destinations: a small amount of material will be crushed and used as a surface treatment for

Plate 1 Simulation interface beta page, nascent phase of development.

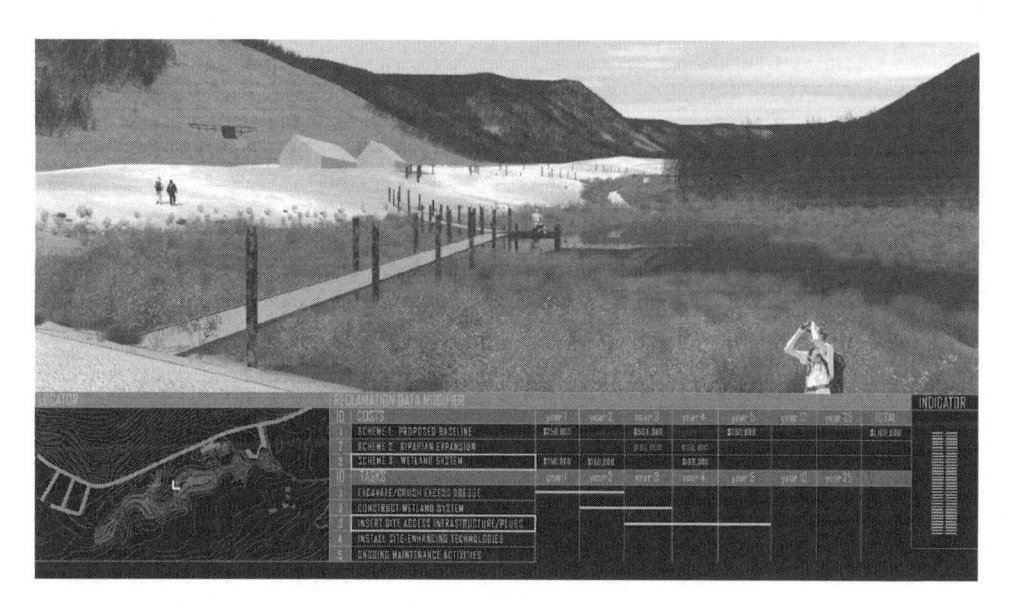

Plate 2 Simulation interface beta page, mature phase of development.

Plate 3
French Gulch valley floor with dredge rock, Breckenridge, Colorado.

Plate 4
Conceptual site plan for new housing development integrated with landscape reclamation, French Gulch.

Plate 5
Proposed site plan. The design integrates recreational use, trails, wetlands, passive water treatment facilities, greenhouses, maintenance facilities, and parking.

Plate 6
Simulation of phasing, program implementation, and ecological maturation of the site over time.

Plate 7
Simulation of seasonal diversity of the fully programmed and mature site over time.

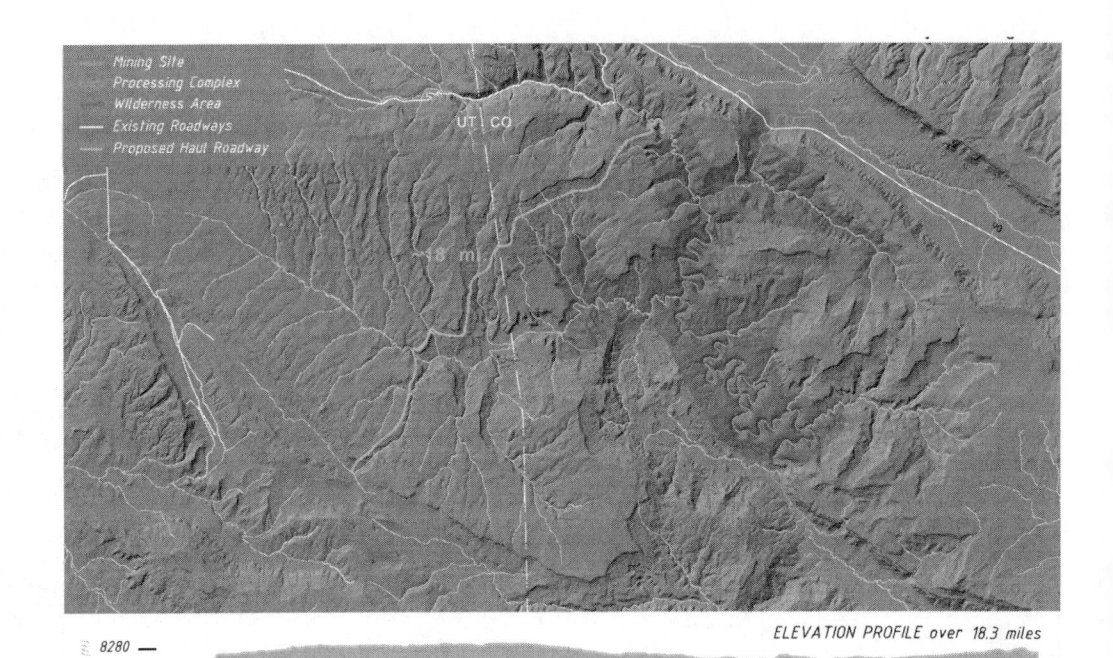

Plate 8 Haul road layout.

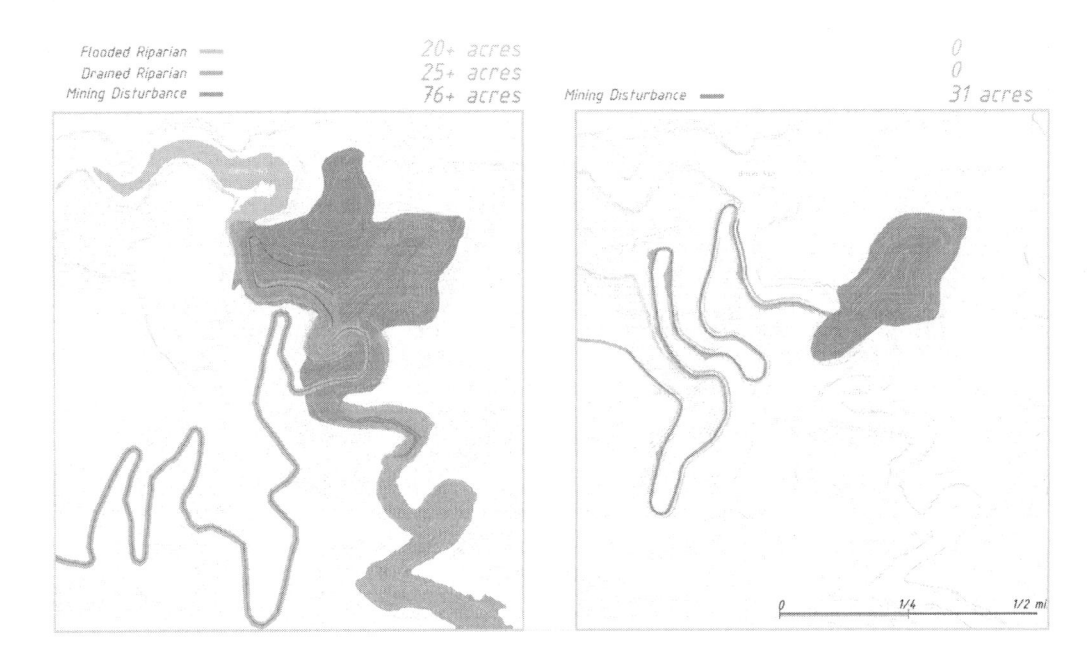

Plate 9 Mining plan disturbance comparison.

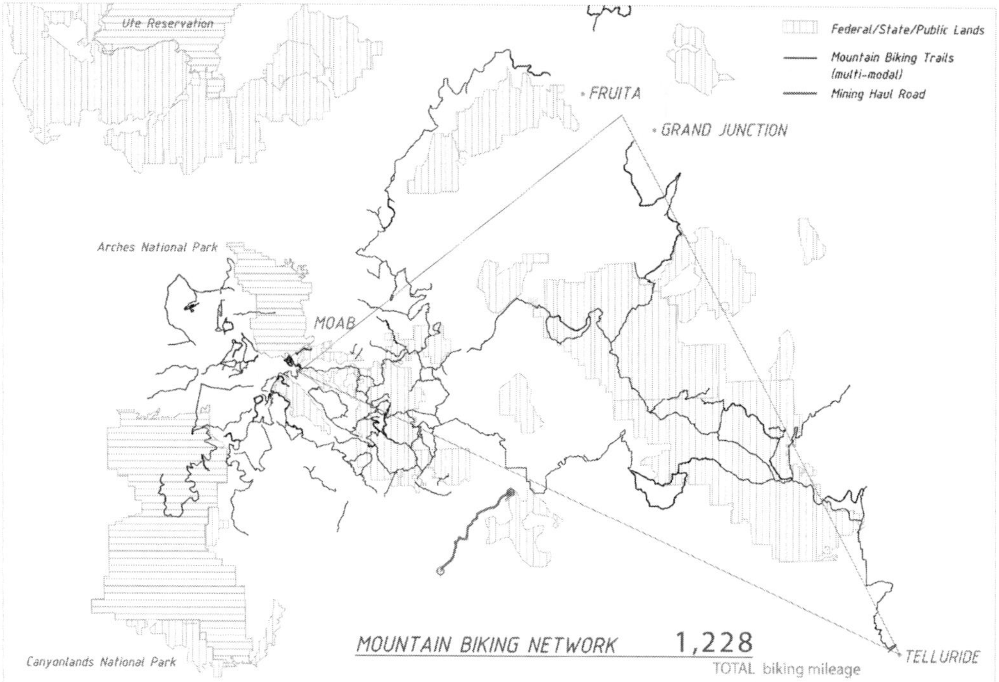

Plate 10 Mapping the regional mountain biking network.

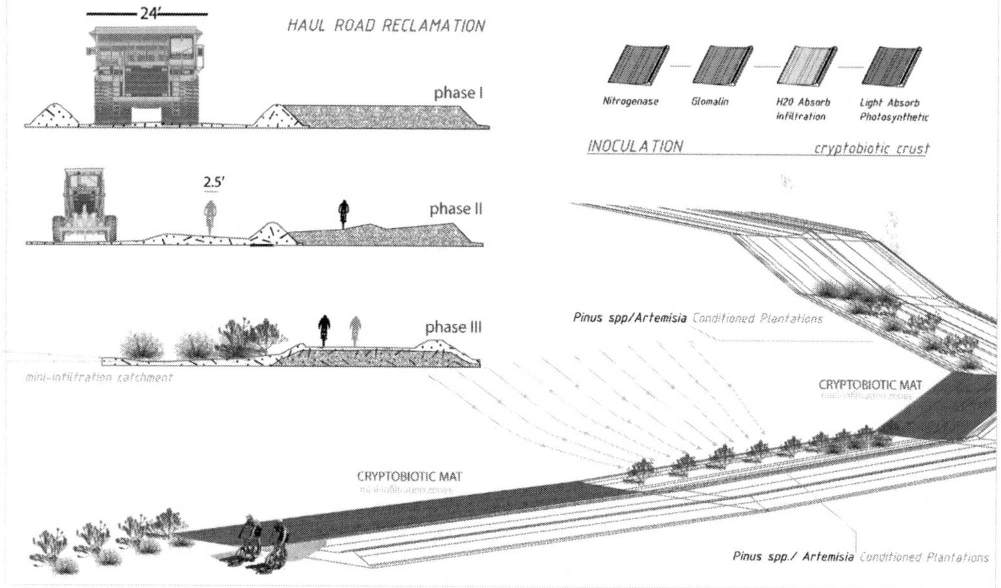

Plate 11 Phased reclamation and reuse of the haul road.

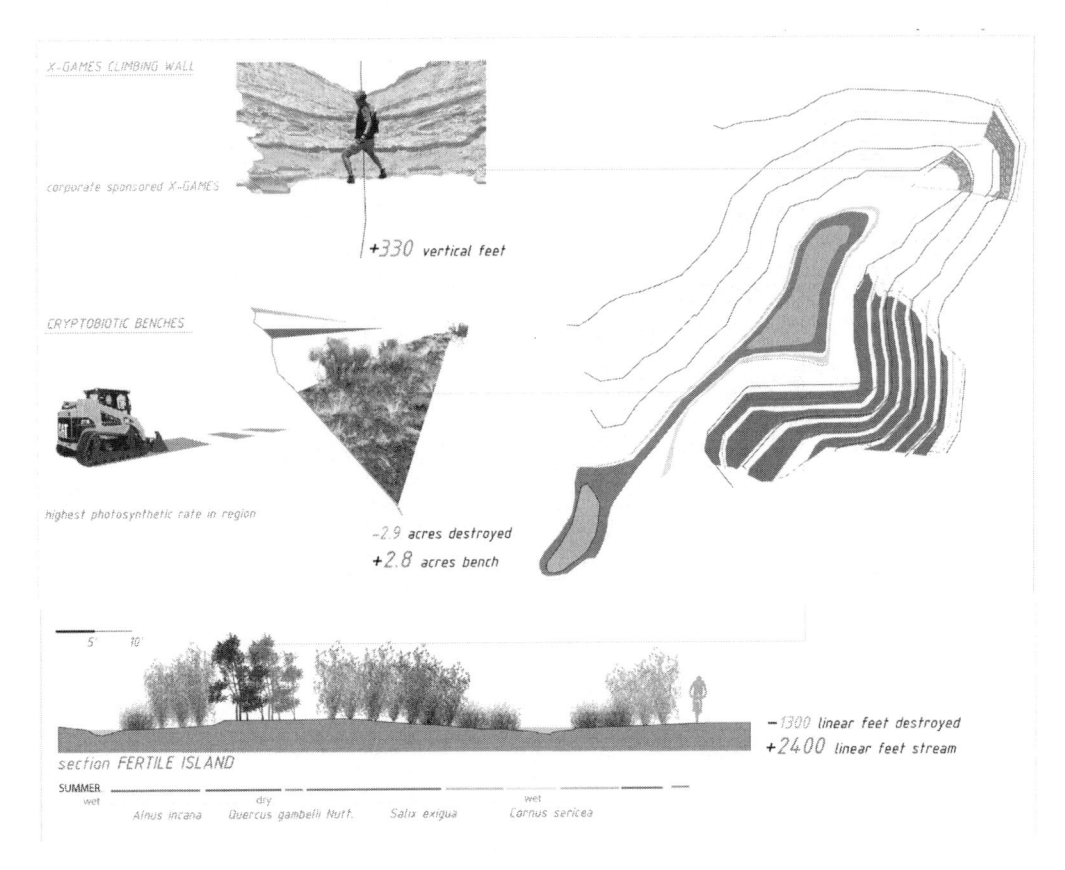

Plate 12 Open-pit re-design.

Plate 13 Implementing recreation programming with final reclamation.

Plate 14 Waste rock piles and tailings along French Gulch access road.
Mine waste like this covers the valley floors along both French Gulch and the Swan River. Gravel companies have periodically removed small amounts of waste rock, but there remains much that needs capping and much that could be safely crushed into a base for trailhead parking lots.

Plate 15 Historic road at French Pass.
This road rises from the Golden Horseshoe, which is the hilly forested area in the middleground, and provides access to other trail systems on the far side of French Pass. Mountain bikers and hikers use this scenic route to enjoy views of the high peaks of Bald Mountain and Mount Guyot.

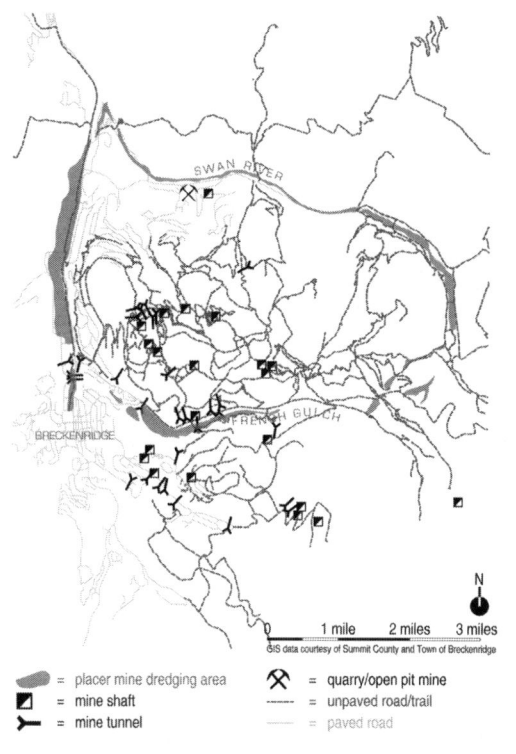

Plate 16 Existing conditions and trail network.
Two valleys bounding the northern and southern sides of the Golden Horseshoe provided access for miners, who also dredged the river bottoms, leaving extensive piles of waste rock. The old mining roads are an unplanned labyrinthine mess now used by recreationalists.

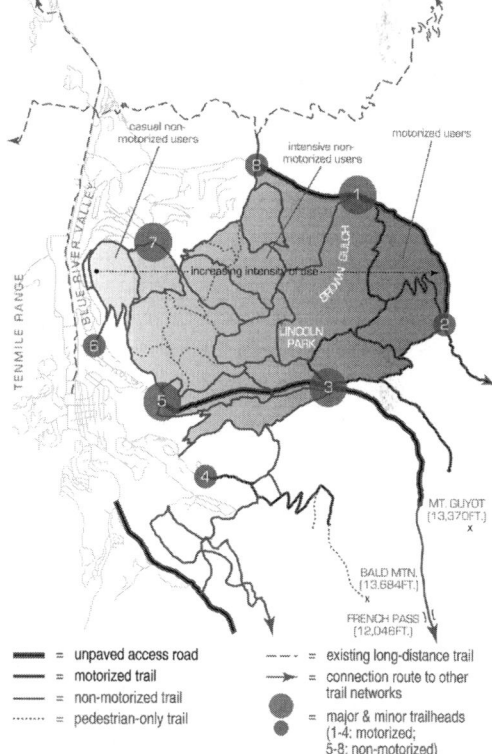

Plate 17 Trail consolidation and "zoning."
Redundant trails are eliminated and new trails are cut where necessary to prevent ecological damage and to create better links between trailheads and within user zones. This map represents only the main arteries within the system. The reconfigured trail network also emphasizes connections to regional networks that enable access to many more miles of trails, particularly for long-distance users, such as ORVs and snowmobiles.

Plate 18 Flume trail initial condition. The surrounding mountains are difficult to see through the lodgepole monoculture. The trail surface is wide and degraded.

Plate 19 Flume trail modified condition (fall). Through selective cutting and aspen succession, the surrounding mountains are revealed as an orienting landmark and the trail's visual interest is enhanced. The trail surface is scarified and restored.

Plate 20 Flume trail modified condition (winter). When the aspens shed their leaves, the trail creates a striking view corridor toward the Tenmile Range, perfect for cross-country skiing at sunrise.

Plate 21 Backcountry gulch trail video rendering (stabilized condition). Three-dimensional computer models enable us to create animated videos that simulate the experience of traveling a trail. In this video (15 sec.: 360 frames), we imagine ourselves as runners on a steep eroded path in the woods.

Plate 22 Backcountry gulch trail video rendering (post-burn condition). After the management interventions depicted in Plates 25–29, we have created a new meadow that offers expansive views and more diverse habitat. In this video (29 sec.: 696 frames), we mountain bike down the switchbacks.

Plate 23 Management evolutions. Initial condition.

Plate 24 Management evolutions. Year 1: Vegetate sides of trail with understory shrubs; construct bridge over stream; restore impaired riparian crossing. (year 1)

Plate 25 Management evolutions. Year 2 or 3: Conduct prescribed burn, using trail and riparian corridor as firebreaks. (years 2–3)

Plate 26 Management evolutions. Year 2 or 3: Replant sides of trail with grass to seed meadow; reestablish riparian vegetation. (years 2–3)

Plate 27 Management evolutions. Years 4 to 8: Close old trail through revegetation; create switchback through meadow; remove burned detritus along new path; allow aspen and wildflowers to colonize disturbed area. (years 4–8)

Plate 28 Section representing trail near mountain wetland. Mountain wetlands are currently rare within the Golden Horseshoe, but a number of areas could (and probably formerly did) support them. Through the creation of treatment ponds or the reintroduction of beavers, we could engineer or restore wetland habitat, thus greatly increasing the area's biodiversity and recreational interest.

Plate 29 Schematic section with animals and habitats. By routing a trail near, but not through, mountain wetlands, we can avoid the human-caused ecological damage now occurring in areas such as Lincoln Park, while exposing visitors to a variety of habitats and animals characteristic of the region, including the tree swallow, moose, trout, beaver, mountain lion, and great horned owl.

the high-access trail areas, while the majority of it will be sold as gravel to cover some of the startup costs. Representatives from Breckenridge estimate upwards of $2 million in profits from the sale of this much rock. After the rock is removed, several areas will open for new terra-forming and reprogramming. A flattened area will be located near the existing tourist mines and will provide parking areas to those entering the backcountry open-space trail system. Other flattened areas will be located along the main roadway to provide the town with staging locations for the stockpiling of soil-making materials, for site-maintenance sheds, and for greenhouses and incubation beds for re-vegetation. These new surfaces can also support additional parking to accommodate the area's increased use as a recreational hub.

Down valley, adjacent to the Wellington Oro neighborhood, trails will follow both the crests of the newly terra-formed dredge-rock lobes and the dips of the newly exhumed riparian zones. In between the neighborhood and the tourist mine, three progressively larger wetland cells will meander through the dredge lobes. Each of these wetland cells will be accessed by a light infrastructure of paths that lifts the visitor off the dredge and over the delicate, emergent ecosystem below. Here the simulations showcase the true power of the digital work flow to represent the complex edges of plant-species assemblages as they interact over time with the reflective water surface (please refer to Plate 6).

The subtleties of the exhumed riparian zones and the wetland cells are elucidated by a detailed examination of their structure and function. With much of the dredge removed to the datum of the intermittent stream, the areas prone to seasonal flooding can be widened to accept more growing medium and its associated biologic material. The widening empowers biological agents to exploit the periodic flooding and drying cycles to accumulate sediment (please refer to Plate 7). Eventually, the germination of seeds from upstream seed banks will occur, rapidly accelerating the colonization by other plant materials. This process could, of course, be catalyzed by the importation of some transplanted vegetation for the collection of sediment. However, this adds cost to the project, and can be evaluated along with the management sequence and work flow with the digital-simulation interface.

The soil and new hydrologic conditions are the foundation for the vegetative population of willows, sedges, and alders – all of which are present in the upstream populations – to perform remediative functions through various nutrient, water, and material recycling. As described earlier, the design strategy is essentially to "disturb the disturbance," or to alter the waste to revitalize the processes that allow ecological regimes to resume activity.

The edge slope of the terra-formed dredge rock is 3:1, enabling both a decent prospect over the evolving wet ecosystem and a negotiable slope down to it. As ecological recovery takes hold, a metal ramp and overlook will be built to provide access and views of the rehabilitated wetlands. Depending on budgetary constraints, further enhancements of the light infrastructure can be made. In the winter, for example, trails could function with minor maintenance by designing an off-the-grid solar-panel system for the metal paths. Such a system would slightly elevate the metal's temperature, melting the snow that periodically falls on it. The wetland cells could be similarly fitted with the same seasonally interactive light infrastructure. This access might help visitors learn firsthand how the wetland slows the water and encourages the growth of plants that collect sediment and filter the toxic metals out of it. In addition, a control structure between each wetland cell prevents non-native trout from invading the rare native cutthroat-trout habitat upstream. Eventually, the cells might also attract insects and other fish populations, as well as their avian predators, thereby renewing the local food web. In total, the design adds two acres of wetland and two and a half acres of riparian corridor, significantly expanding those critical and endangered Rocky Mountain habitats.

Coda

Digital simulation is a valuable tool for building consensus among stakeholder groups in reclamation-design projects. Simultaneously, users can experience designed space and analyze the construction phasing and cost implications that produce the new land uses. Ultimately, digital simulation provides a platform for the users, designers, and altered landscapes to meet on common ground. The French Gulch Simulation Project and its unique abandoned quality and ecological challenges occasioned an expansion of thought about digital simulation and the evolutionary roles that can be embedded in representational formats. Only in such an interfacing environment can reclamation design fulfill its role at the nexus of the human–land relationship: no longer out of sight, out of mind.

Notes

1 Victor Ketellapper provides an extensive background of this site's regulatory history in "The Wellington Oro Mine-Site Cleanup" (see Chapter Eight). The research undertaken by P-REX on the French Gulch/Wellington Oro site was funded by grants from the Town of Breckenridge and the U.S. EPA's Office of Land Revitalization and Superfund Program.
2 The stakeholders meeting took place in Breckenridge, Colorado, at the Town Hall, where representatives from the town's Planning, Trails and Open Space Office, Town Manager's Office, Summit County, and the U.S. EPA Superfund Program were in attendance.

Bibliography

Berger, A. (2002) *Reclaiming the American West*, New York: Princeton Architectural Press.

Johnson, M., Cooke, J., and Stevenson, J. (1994) 'Revegetation of metalliferous wastes and land after metal mining,' in R. Hester (ed.) *Mining and Its Environmental Impact*, Cambridge, UK: Royal Society of Chemistry.

National Park Service (NPS) (2002) *Great Basin National Park*. Online. www.nps.gov/archive/grba/bats.htm (accessed November 8, 2006).

Power, T. (1996) *Lost Landscapes and Failed Economies*, Washington, D.C.: Island Press.

United States Department of Agriculture (USDA), Natural Resources Conservation Service (NRCS) (2002) *PLANTS Database*. Online. http://plants.usda.gov/java/profile?symbol=SAEX (accessed January 11, 2007).

Chapter 13
Open-pit opportunities:
pre-mine design strategies

Alan Berger and Case Brown

Background and context

Open-pit mining for metals causes an enormous amount of physical alteration in the landscape. Haul roads, pits, waste-rock impoundments, and processing facilities are a few of the myriad components that mine operators must build during the metal-mining process. The economic costs of this mining include extracting and processing the target minerals and reclaiming the landscape to a desired state or condition. In most US states, a reclamation plan and secured bonding are required prior to the issuance of a mining permit. The focus of the mining industry is, in general, to minimize the projected reclamation costs of a particular operation while meeting the required environmental standards and approvals to receive a mining permit. As a result of this business model, mine operators often overlook the hidden economic, programmatic, and environmental potentials of landscape reclamation. The following study describes a unique case, conducted by the Project for Reclamation Excellence (P-REX), for which the commissioning private mine operator wanted to explore the pre-mining landscape-reclamation-design potentials of an open-pit copper complex on the Utah-Colorado border.

The mine's owner and operator had plans for developing an open-pit copper mine that straddled a canyon in the Colorado Plateau geographic region, taking advantage of an existing processing facility in the region, some eighteen miles away (see Plate 8). Processing facilities are usually the most expensive components of metal mining; they are the locations where ore is mechanically or chemically turned into valuable pure metals. In most mines, this facility is located near the pit in order to reduce the transportation costs associated with hauling millions of tons of ore. After a cost–benefit analysis, the mine stakeholders determined that it would be less expensive to reuse the existing processing plant, even though it would mean constructing eighteen miles of haul road from the new pit.

Given the unusual distance between the pit and the processing plant, P-REX researched alternative uses for the haul road that would provide social, economic, or environmental benefits as the result of reclamation. The scale of the haul road highlights the prescient need to develop design strategies for integrating it back into the landscape after mining ceases. Additionally, the road consists of fifty-foot-wide ramps connecting the haul road to the rim of the pit. The total size of the disturbance, including haul road and ramps, is one hundred and eleven acres, mostly in a linear configuration. This breadth is significantly larger than the active mine pit itself, which is seventy-six acres (see Plate 9).

The mine operator provided P-REX with unrestricted access to its pre-mine permit plans, economic data, and cost–benefit analysis, in addition to allowing the Project to tour the local site and conduct its own site analysis. Instead of focusing solely on the mine owner's site plan, P-REX looked extensively at the regional landscape systems that interacted with the site and which would be impacted by the mine's operations. Hence, much of the work became framed by "before/after" evaluations of the landscape: how could the mine's largest components (the haul road, ramps, and open pit) be transformed into landscape assets for regional plant associations and human and animal inhabitation? And how could the site be linked into larger recreation networks in the region?

The importance of riparian corridors in arid regions

After conducting its site analysis, P-REX discovered that the proposed haul-road location straddled a vibrant, intact riparian canyon. It has been estimated that approximately 1 percent of the riparian habitat of the American West remains (Knopf et al. 1988), yet 75 to 80 percent of all species in these arid ecosystems depend on that habitat (Gillis 1991). The complex web of species that are adapted to, and even dependent on, the riparian system designates it as a resource worth maintaining. The riparian corridor reigns as the central ecological, and potentially economic, factor in arid regions. If the costs of replacing such an area were factored into the cost–benefit analysis of the mine (which the law does not require), it would be projected that the mine would likely lose money. The viable alternatives that would allow for both the extraction and protection of the riparian corridor were therefore further explored as primary options.

Haul road to mountain biking

In remote areas with rolling or mountainous terrain, it is common practice for mines to use valley areas as waste-rock impoundments, or dumps. While this practice does save expense in the short term (there is less hauling of waste uphill, and gravity is used to move excess material away from the site), it usually causes long-term environmental problems for the water bodies flowing through the valley cavity. Instead of compromising the entire riparian corridor with waste-rock impoundments, as the existing plan proposed, P-REX designed a scheme for the waste rock to be utilized as fill material along the entire eighteen-mile haul road. The resultant road width would become significantly wider with the excess fill – almost double what the original mine plans called for – while the riparian corridor would be left intact (see Plate 9). The construction of the haul road would inevitably act as a physical, linear barrier across the existing landscape systems. Further study concluded that doubling the road's width added little extraneous disturbance to the mesa's piñon-juniper ecosystem, while potentially salvaging the more ecologically valuable riparian system. The widened road, moreover, would allow for more flexible post-mine land uses, as well as accessibility to the open pit if it were to be programmed for new uses in the future.

Another discovery that P-REX made while conducting its regional landscape analysis is that Moab, Utah, serves as the national "nerve center" of mountain-biking activity. There is a dense network of mountain-biking trails in the area, which link Moab to other small population centers. Many towns and counties in the region support the trail network and are increasingly relying on recreation-based economic opportunities. This economy is seasonal and service-based, allowing the towns to utilize the vast landscape amenities with minimum public investment for improvements. In order to see how the mine site could fit into or extend the trail network, P-REX collected all of the regional trail maps and geographically referenced and re-drafted them to reveal relationships between the biking trails and the mine site (see Plate 10). A triangular network is developing between Moab, Fruita/Grand Junction, and Telluride, Colorado. The haul road and mining site fall within this triangulated area, allowing it to be easily linked into the existing system of trails.

One potential reclamation design for the haul road is to use its extra width for modal separation of future recreation users. The haul road could be developed into a topographically diverse biking trail for multiple skill levels and various modes of recreation (hikers, bikers, and vehicles). This new topography would require some back-blading and incur some costs, but it could pay for itself over a short period of time through user fees, which are not uncommon for special trails and event-based trail usage, such as races (see Plate 11). The unique reclamation design of the haul road would have the potential to transform the mine site into a major mountain-biking destination.

Designed experimentation: pit, surface, and haul-road ecology reexamined

Another reclamation design opportunity arises in the haul road's ability to collect and redistribute surface water. Hydrological gradients control many of the ecological functions in the arid Colorado Plateau landscape, including vegetation growth and nutrient cycling. With the aid of properly graded infiltration zones along the upslope side of the haul road, water could collect and create micro zones for supporting diverse vegetation (see Plate 11). Cryptobiotic soil – a symbiotic, crustal mat that thrives in extreme aridity and plays an important role in the drought-resistant properties of plant and animal communities – could be added to the infiltration zones to stabilize the soil. Cryptobiotic soil is capable of absorbing the infrequent rains in the thalli of lichens and of fixing nitrogen for use by the various photosynthetic organisms living in the medium. The symbiotic system would decrease erosion and, by thus maintaining the structural integrity of the soil in an agglutinated mat, will/would hold essential minerals and nutrients for other kinds of organisms. The large, linear surface area of the haul road is the perfect type of site for performing reclamation experiments that will enable a better understanding of cryptobiotic soil properties and the disturbance dynamics of reclamation ecology. There is a particular interest in conducting scientific experiments in the arid West, where the largest concentration of abandoned mining sites is located. The largest land-related federal agencies of the American West (the U.S. Environmental Protection Agency, the U.S. Fish and Wildlife Service, the U.S. Bureau of Land Management, and the U.S. Forest Service), along with various state-level entities, could be allocated sections of the eighteen-mile haul road to conduct reclamation experiments.

The mining pit itself possesses several attributes that could be transformed into recreational venues. First, the back highwall of the pit could be designed with climbing walls. With some stabilization engineering, it could be one of the first man-made/natural-rock-formation climbing walls in the world. It could serve as an arena for the rock-climbing sector of recreational users and generate funds by staging events and competitions sponsored by private organizations. Extreme sports are of particular programmatic relevance here, owing to their popularity among young adults, particularly in the Intermountain West (see Plate 12). Second, with some minor recontouring on the northwest-facing slope (chosen for its increased moisture and for its inaccessibility, which will keep visitors from interfering with the reclamation), flatter zones could be created in order to re-inoculate the soil with cryptobiotic organisms, catalyzing the process of re-vegetation.

Mining operations in the bottom of the pit will destroy much of the existent riparian condition and leave some areas barren. The P-REX design strategy for reclaiming the bottom of the pit is to draw water from the existing creek during seasonal flooding periods. The new water will deposit fine sediment from upriver sources and replenish the barren soil. There are native plant species, such as alder (*Alnus incana*), that are adapted to such harsh conditions and perform the function of stabilizing stream banks (see Plate 13). This species, along with willow (*Salix spp.*) and red-oiser dogwood (*Cornus sericea*), is capable of colonizing flooded, unstabilized stream banks. The re-vegetation strategy also includes cost-effective bare-root planting techniques, which should produce the base for a vibrant, reclaimed riparian corridor. Lastly, with the insertion of a kind of raised, fertile island in the middle of the pit, point bars would likely develop, helping collect the sediment. If these measures are taken, the pit design could actually result in more riparian habitat than was present before the mining operation. All of the stakeholders ultimately prefer such an outcome to the habitat eradication that would occur if the waste rock were to be dumped into the canyon.

Coda

Recreation, reclamation, and the generation of new ecological habitats are possible and capable of coexisting with a mined landscape. This project stands as a possible set of strategies for achieving an ecologically and economically productive design in a mined landscape. Taking advantage of these strategies to evaluate the deeper economic impact on landscape-based tourism and functioning ecologies enables novel approaches to extractive landscape reclamation. By seeking to amplify, test, and diversify pre-mine landscapes via manipulation of actual mining procedures, designers, architects, and mine stakeholders can turn pits and scars into invented habitats and recreation beacons.

Bibliography

Gillis, A. (1991) 'Should cows chew cheatgrass on common lands?,' *BioScience* 41: 668–75.

Knopf, F., Johnson, R., Rich, T., Samson, F., and Szaro, R. (1988) 'Conservation of riparian ecosystems in the United States,' *Wilson Bulletin* 100: 272–84.

Chapter 14
Reclaiming the woods:
trail strategies for the Golden Horseshoe's historic mining roads

Alan Berger and Bart Lounsbury

Abandoned mines can appear dramatic, as streams flow orange with colored sediment and disfigured tailings piles and open pits create a surreal topography. Anyone who happens upon these areas knows immediately that something is awry. Most reclamation efforts focus on ameliorating the environmental problems associated with these visually striking consequences of mining: heavy-metal exposure, acid mine drainage, groundwater contamination, and so on. From an ecological perspective, such efforts are undoubtedly the most important, but in many places the effects of mining extend well beyond the polluted rivers and rocky mounds that catch the casual observer's eye.

Across the United States and other parts of the world, speculators and mining companies have wantonly blazed roads into the wilds, sometimes prospecting for new lodes, sometimes seeking wood for fuel and construction. Once the land or the miners' own resources were exhausted, they moved on, typically leaving these sites denuded of vegetation and crisscrossed by the labyrinthine networks of vehicle tracks that had been used to haul away timber and minerals. In Breckenridge, Colorado, remnants of such activity are omnipresent (see Plates 14 and 15). The forests of the heavily mining-impacted Golden Horseshoe, an 8,600-acre area northeast of the resort town, are mostly a monoculture of lodgepole pines, traversed by a confusing hodgepodge of old mining roads. Recreationists have discovered the Golden Horseshoe, but only a few hardy souls currently manage to penetrate the web of trails that once carried miners to prospects and timber stands. Despite the area's potential for recreation uses, a lack of trail-system planning and forest management has resulted in a landscape with little ecological or experiential diversity. What follows is a comprehensive planning and design strategy for the reclamation and reprogramming of the Golden Horseshoe's abandoned mining roads into a regional recreational trail system.

Navigating the spaghetti

Geographic information systems (GIS) data taken from the Golden Horseshoe reveal over one hundred miles of abandoned mining roads and associated trails (see Plate 16). In some places, nondescript trails parallel one another a mere fifty to one hundred feet apart. In other places, trails travel directly through sensitive riparian areas and climb steep grades that are prone to erosion. Except for the westernmost portion of the Golden Horseshoe, where it drops into Breckenridge, there are no signs to direct visitors and very few landmarks to orient the wayward traveler. Even the most experienced local hikers and backwoodsmen remark that finding one's way through the Golden Horseshoe requires a little guessing and a lot of luck. The area's trails desperately need consolidation and reconfiguration in order to become a recreational attraction and to reduce the negative environmental impacts of increased human use. The planning and design goal of this project is to transform the Golden Horseshoe's roads and trails into a regional recreation system while reclaiming biological and ecological diversity in the forest. The project is thus both a forest-management strategy and a recreation plan.

Trail-system zoning and access

Miners would probably have dropped their gold nuggets in surprise if someone had told them that the roads they were cutting out of the earth would someday evolve into recreational trails for people on mountain bikes, cross-country skis, and all-terrain vehicles. Most of the paths in the Golden Horseshoe were created for mining and logging, not for outdoor recreation; consequently, these paths' suitability for recreational use is limited. As the trail network currently exists, any type of user can access the backcountry from a number of unmarked points along the major roads in the valleys. This pattern of unregulated access prevents public authorities from having any control over who uses the trails or where and how they are used, which leads to conflicts when different types of users travel the same paths. Most frequently, the conflicts arise when the noise and speed that characterize different forms of recreation are incompatible (e.g. motorized vs non-motorized users, mountain bikers vs hikers).

To mitigate such conflicts, the project utilizes the Golden Horseshoe's natural topography to create trail-system "zoning" (see Plate 17). Within this proposed system, casual non-motorized users would access the historic flume trails, which already weave through the backyards of luxury homes, close to town, where the Golden Horseshoe rises from the Blue River Valley in the west. Several hundred vertical feet above Breckenridge, where public land begins in the forests, the Golden Horseshoe flattens out. Here the more intensive non-motorized users would take over from the dog walkers and after-work runners on the lower trails. Several miles to the east, the steep ravine of Brown Gulch and the wetlands of Lincoln Park would separate non-motorized users from vehicles such as dirt bikes and snowmobiles. These motorized users would have the run of this easternmost portion of the Golden Horseshoe, where a downslope toward the east would attenuate the vehicular noise that might otherwise disturb non-motorized users in the western parts of the system. Here trails would also connect to existing cross-mountain routes capable of accommodating the long ranges of motorized users, who would access the backcountry via the main roads in the two valleys that inscribe the region's "horseshoe" shape.

The key to establishing this trail-zoning system is to restrict and control recreational access. This requires the closing of numerous entry points into the Golden Horseshoe and concentrating access at strategically placed trailheads. Several such trailheads already exist and simply need expansion and redesign to become more obvious to users. The redesign of these access points could coincide with various mine-remediation activities – the crushed waste rock that has been removed from the riverbeds, for instance, could provide a base for parking lots, which might also function as caps on tailings piles. Additionally, trailhead parking lots could accommodate visitors to nearby mining sites that have been remediated and specifically designed for humans to experience and (within sensible limits) interact with the reclamation process. Some of the trails that depart from these trailheads could also serve an important interpretive purpose by following and describing the historical practice of mining in the area (e.g. explaining the flume trails as mechanisms of water capture and delivery or the timber harvest as a means of fueling smelters).

Individual trail enhancements and graphic modeling

Even with inviting trailheads and the appropriate separation of incompatible uses, no one will enjoy a backcountry recreation network if the trails themselves are unexciting or severely degraded. The Golden Horseshoe's current network contains many old mining roads that are of relatively limited aesthetic and recreational interest, or that are eroding or impairing sensitive ecological areas. Adopting the proposed trail-design and maintenance strategies requires actions foreign to most trail planners, but their adoption would significantly improve both the users' experience and the ecological characteristics of the area.

In light of the scale of the Golden Horseshoe project and the diversity of the stakeholders involved in the planning and decision-making processes (the U.S. Forest Service, the Town of Breckenridge, Summit County, and homeowners associations, among others), the project brings its zoning, design, and maintenance strategies to life with computer models of trail segments, which can be animated and altered as though the viewer were experiencing the trail's evolution and management over time. The usefulness of this digital-simulation design protocol is that, rather than producing a single plan view or analytical planning document, it gives stakeholders the ability to manipulate the trails' evolution through their own management interventions, as they trigger or redirect ecological processes. A few of these strategies are described below and depicted in the accompanying images.

View-corridor enhancement

One of the Golden Horseshoe's primary assets is its proximity to numerous rugged and treeless high-elevation peaks. Views of these peaks and other ridgelines are now blocked by vegetation on the existing roads and trails. Opening up such vistas through the management interventions described below would enhance the trail users' experience and provide landmarks for directional orientation (see Plates 18, 19, and 20).

Selective tree cutting and prescribed burns

The trail system's managers could create view corridors through selective cutting and prescribed burns, which, in addition to opening up viewsheds, would provide ecological benefits to the forest. Selective cutting could reduce the danger of wildfires and potentially counter the effects of the pine-beetle infestation that is currently sweeping through Colorado. Prescribed burning would serve the same purposes, while also catalyzing ecological succession in upland meadows and aspen stands. The area's lodgepole pines, moreover, depend on fire to release their seeds and regenerate (see Plates 21–27).

Species diversity and trail consolidation

The various biotopes fostered by cutting and burning should permit more species to reside in the Golden Horseshoe and would therefore create exciting new wildlife-viewing opportunities for users as they travel through the area (see Plates 28 and 29). These trail-management strategies also provide planners with an opportunity to close off or redirect routes that have detrimental environmental impacts, such as erosion and water diversion from polluting mine adits.

Coda

This trail-redesign project uses digital simulation to demonstrate different methods of reclaiming a broad spectrum of abandoned mining roads and trails in Colorado's Golden Horseshoe. By envisioning reclamation through design protocols that are sensitive to diverse user groups, each of which has a vested interest in the future of the regional landscape, the project allows better reclamation planning and design to occur. Simultaneously managing vast forest areas and developing a strategy that acknowledges the legacy of mining through trail reuse and integration, this scheme for the Golden Horseshoe will, if implemented, benefit both humans and the ecosystems they have altered.

Chapter 15
Real-time coal mining and reclamation:
the Office of Surface Mining's Technical Innovation and Professional Services program

Billie E. Clark, Jr.

There are four primary services that the Office of Surface Mining's Technical Innovation and Professional Services (TIPS) program provides. It delivers software. It delivers hardware. It trains users in how to use that software. And it provides specific technical assistance to the users of that software.

Who are the program's customers? State governments enforcing the Federal coal surface mining and reclamation law within their state boundaries are the primary customers.[1] In addition, the program provides technical support to tribal regulatory programs and Office of Surface Mining (OSM) offices administering the Federal law where the states do not (e.g. OSM enforces the Federal law on Indian lands). Before a coal-mining company can start mining, it has to secure approval from the state government or OSM by submitting a comprehensive mining and reclamation plan to the regulatory authority. The mining and reclamation plan is then reviewed and approved if it is determined that it is in compliance with the Federal law. So state governments, tribal governments, and OSM offices are TIPS' customers. There are, in fact, about seven hundred TIPS customers throughout the country, from Maryland to Alaska, using twenty-six different TIPS-provided scientific/engineering software applications on their computers. Environmental Systems Research Institute (ESRI) is the primary Geographic Information System (GIS) tool. Other software applications available through TIPS include Autodesk's AutoCAD, Dynamic Graphics, Inc. (DGI) EarthVision three-dimensional modeling software, and a whole suite of hydrology software packages. These are the applications used throughout the country by engineers, hydrologists, scientists, geologists, archaeologists, biologists, and ecologists to review coal mine- permit applications and assess field compliance with the approved permit applications. Comprehensive training programs must, of course, accompany the implementation of the software. Implementation of the training program is accomplished by three methods: (1) instructor-based training at regional training centers, (2) web-based e-Training, and (3) project specific on-the-job training.

There are four emerging technology areas of the TIPS program covered in this chapter: GIS, mobile computing in the field, remote sensing, and three-dimensional modeling. Detailed examples are provided below to offer a sense of how software packages are used in the process of reviewing mine-permit applications.

GIS

GIS is essential software that OSM began to use in the mid-1980s, and remains a very powerful tool for the mine-site inspectors and technical reviewers. At the McKinley Mine in New Mexico, for example, OSM has used a software package called ArcGIS (GIS application from ESRI), which includes different information layers such as the permit boundary, topsoil distribution areas, and where the mine is authorized to conduct activities. At this particular mine site (see Figure 15.1), an inspector used the software in the field to gather information related to the topsoil-distribution plan (e.g. topsoil thickness, GPS location). The company is required to replace the topsoil on the ground after mining in accordance with the approved permit application, and, through the GIS application, the inspectors were able to input information for later review to determine compliance.

The Black Mesa Mine in northeastern Arizona is one of the largest coal mines in the world, covering about a hundred square miles. One finds here a patchwork of permit boundaries, jurisdictional boundaries, and special use areas that necessitate application of different regulations (see Figure 15.2). Using the GIS software, an inspector can navigate this patchwork of information to determine which permit or regulatory requirements are in effect on any particular location in the field.

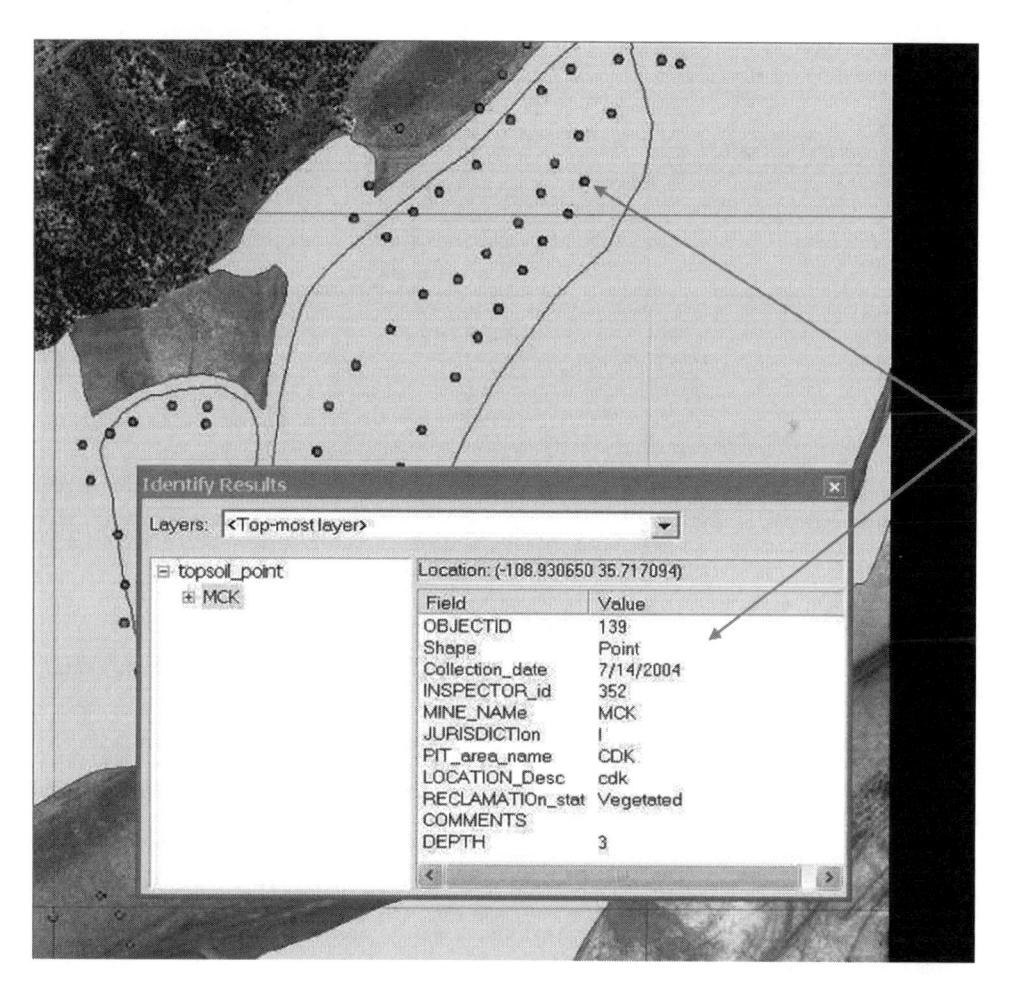

Figure 15.1 In the field, software used to construct a topsoil distribution plan during reclamation.

Billie E. Clark, Jr.

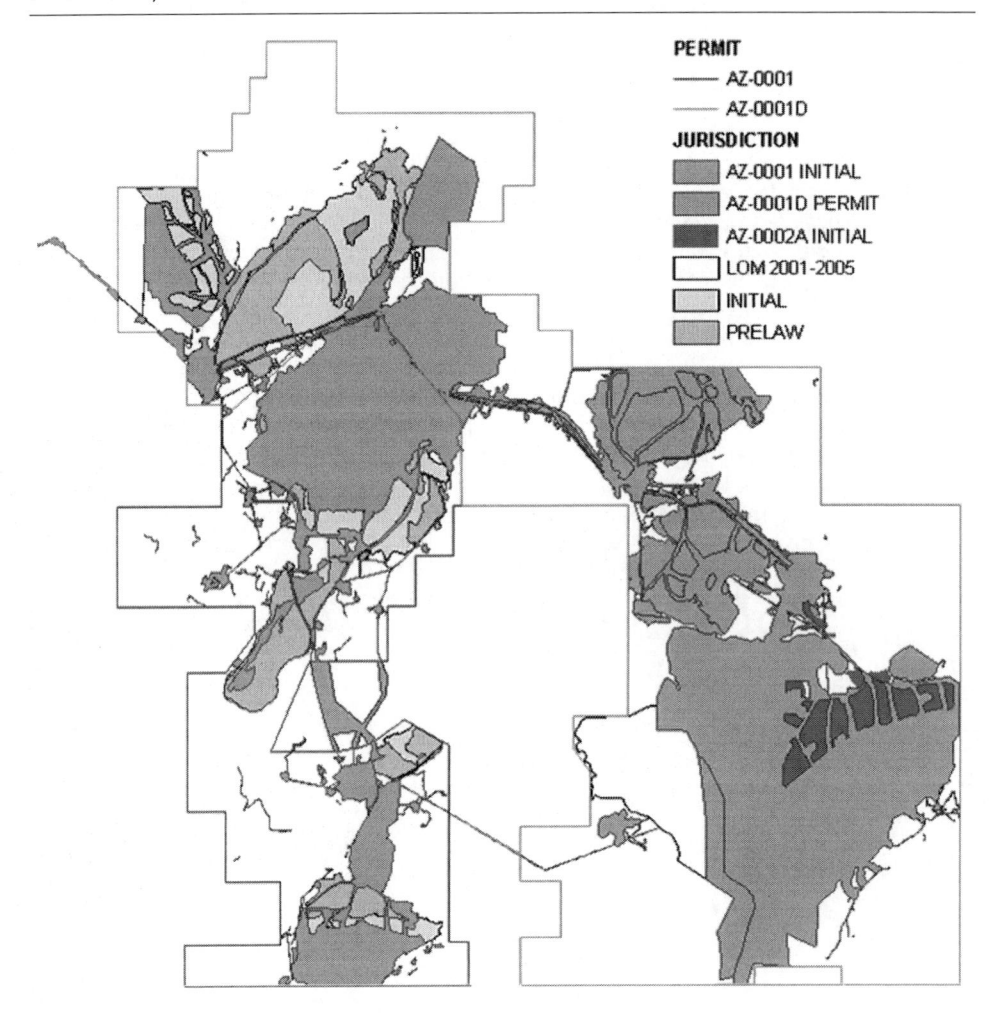

Figure 15.2 Software can be used to help field inspectors distinguish land jurisdictions and permit area boundaries.

Another example of GIS's usefulness is on a coal mine in Wyoming, where a reclamation bond has been secured. Mine operators are required to post a bond (e.g. surety, self-bond) to ensure that, if the mine were to close at any time, the regulatory authority would be able to collect the money specified by the bond and reclaim the site according to the Federal coal surface mining and reclamation law. In Wyoming, there is a comprehensive five-phase process for release of bonds, and, in the case of this particular mine, GIS is helping Wyoming track the status of the bond.

A final example presents an inspector's use of a combination of layers to determine compliance (see Figures 15.3A and 15.3B). When an inspector goes to a mine site to evaluate compliance, he or she brings the approved permit along. Information about drill holes and drainage patterns, to name just two layers, is critical in determining compliance not only with Federal coal surface mining and reclamation law, but also, more specifically, with the particular permit requirements that the inspector is examining.

The OSM's TIPS program

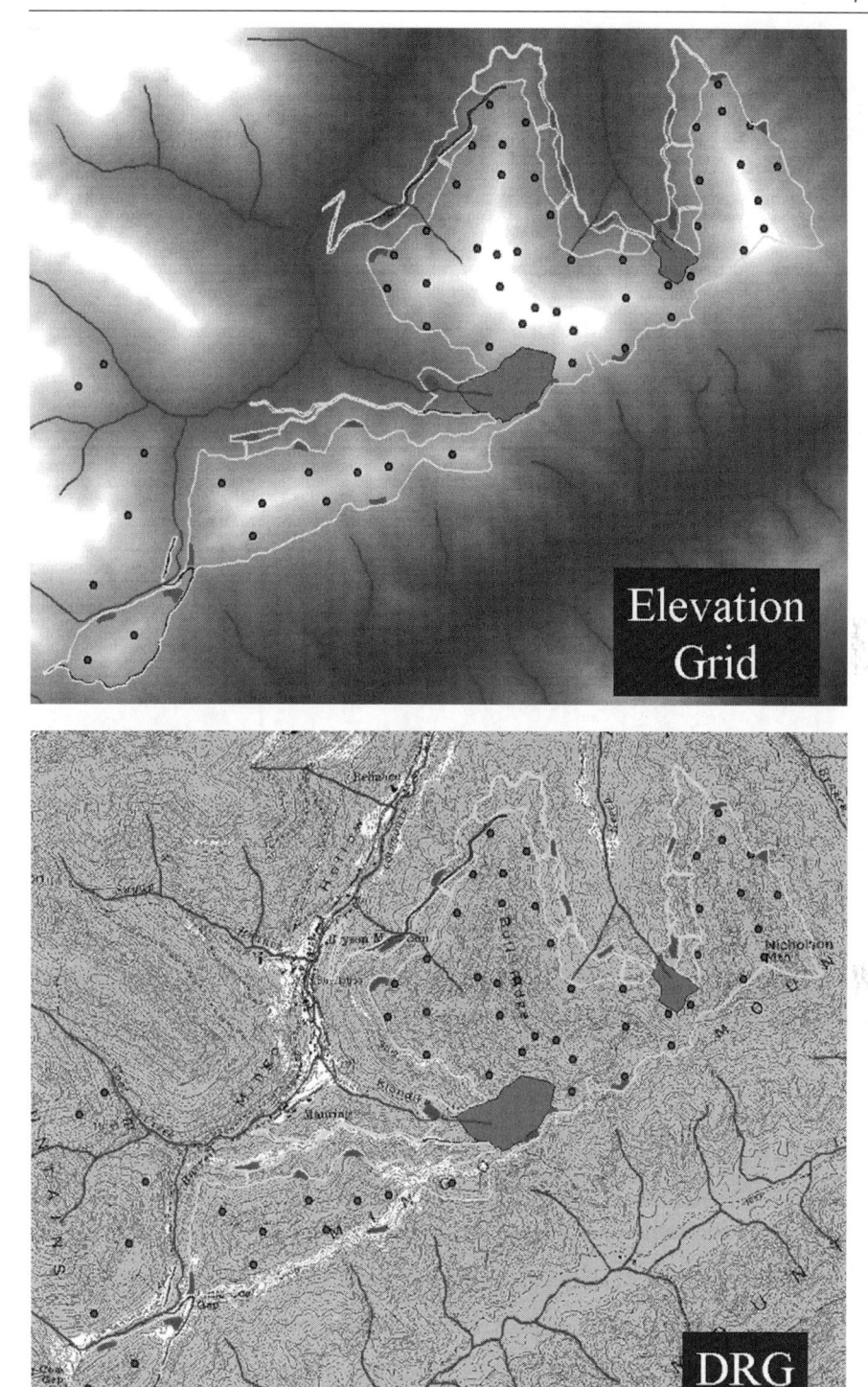

Figures 15.3A and 15.3B Drill hole and drainage compliance map layers can be activated in the field to provide real-time information for reclamation inspections.

135

Mobile computing

Mobile computing is an emerging and exciting part of the TIPS program. Figure 15.4 illustrates the field use of a "tablet" commonly being used by inspectors and field personnel. The Fujitsutablet, with an enabled GPS unit, is just one example; it provides everything an inspector needs in the field: information, maps, even the voluminous permit application.

OSM conducted a comprehensive study over the last few years and is now recommending mobile-computing devices to states, tribes, and OSM offices. Some hardware devices are purchased by the TIPS program to help the states kick-start their fieldwork. At the Wyoming mine mentioned above, the inspector used one of the devices to assess the redistribution of the topsoil. After determining the depth of the topsoil with his own hands and a shovel, he used his GPS unit to record the locations of the topsoil-distribution areas, brought it back to the office, and displayed it on the GIS. He was then able to determine compliance according to the precise amount of topsoil specified in the permit. At another site, photos, GPS, and GIS were used in combination, and the inspector discovered a disturbance boundary where the operator conducted activities outside the permit boundary. Consequently, a notice of violation (NOV) was issued. The operator quickly reclaimed the site.

At a site in Oklahoma, OSM used a mobile-computing device, a fun little all-terrain vehicle, and a GPS unit to collect data to create a new reclamation design. As the all-terrain vehicle traversed the site, the inspectors took continuous GPS readings. They ended up with a topographical map of the entire 240-acre area, and then used that data to help design the reclamation plan. As one can see, mobile computing enables the gathering of data quickly, easily, and accurately.

Figure 15.4 The Fujitsutablet, with an enabled GPS unit, being used in the field.

Remote sensing

OSM is also using remote-sensing technology and the GIS layers to determine compliance with approved permit applications. A few years ago, OSM hired a remote-sensing specialist for the first time. One of his initial jobs was to show the usefulness of remote sensing at a coal mine in the state of Washington. After the image was received and processed, staff immediately recognized a violation of mining activities outside the permit boundary. The inspector had not caught the violation in the field, but it was eventually caught by means of the imaging technology. An NOV was issued, and the operator worked with the land owner to correct the problem.

Returning once again to the McKinley Mine in New Mexico, an inspector used IKONOS satellite images to assess compliance. Various information layers including the high-wall, as well as rough grading, topsoil placement, and re-vegetation areas, were identified by means of remote sensing technologies. Using this information and the approved permit application the inspector was able to conclude that the operator was in compliance with contemporaneous requirements.

Additionally, an interesting problem presented itself at a mine in Arizona. The Federal coal surface mining and reclamation law requires adequate drainage of reclaimed sites, except for approved impoundments. In one reclaimed pit area at the Arizona mine a substantial amount of water was trapped, in spite of the site design's drainage provisions. OSM used remote sensing technology to produce a three-dimensional image that could be interpreted quickly and easily. When OSM met with the operator to discuss the issue, there was no discussion or debate; it was obvious that the drainage in the pit area was not adequate. The imaging technology helped OSM and the operator to work together comfortably to resolve the issue.

At a mine site in the anthracite region of Pennsylvania – a joint project by the State of Pennsylvania and OSM – it was determined that the reclamation bond for the site was not enough to adequately backfill and reclaim a large area. A combination of aerial photographs and ERDAS IMAGINE software allowed Pennsylvania and OSM to easily calculate earth moving costs. Based on a detailed report using these calculations, and derived maps and images, the coal mine was required to post additional bond. The resulting reclamation bond increased from $5.2 million to $13.2 million. The remote-sensing technology helped the parties involved resolve their differences.

The State of Wyoming, which alone produces about half the country's coal, asked OSM for help in a pilot project to determine better ways to track and monitor coal reclamation bond status. A pilot project agreement was struck among OSM, the State of Wyoming, and the Peabody Coal Company (which operates several of the state's mines). Using several of the technologies at OSM's disposal – the ESRI, the GPS units, and remote sensing – a pilot GIS was developed to track the bond status of every acre of two Peabody mines. As the project reaches its conclusion, the State of Wyoming has initiated plans to implement the GIS protocol statewide.

As a final note on remote sensing, the national TIPS team is looking into adding a geomorphic software application, Carlson Natural Regrade with GeoFluv, to its suite of software programs. This software assists in creating a landscape design that mimics the functions and aesthetics of the natural landscape that would naturally evolve over time, resulting in a more stable hydrologic equilibrium.

Modeling

Modeling is the fourth technology area of the TIPS program. The particular modeling software that OSM deploys is a package called EarthVision, by Dynamics Graphics, Inc., which is used by petroleum companies, geologists, and mining companies. In reviewing the permit application for compliance with surface-mining law at the McKinley Mine in New Mexico, OSM used EarthVision modeling techniques to predict if final topsoil and root zone placement would be adequate.

The permit application for the McKinley Mine requires, for example, that the root zone (the top four feet below the surface) has a pH within a particular range and that the selenium be less than 0.8 parts per million. The permit application must demonstrate that these requirements can be met before the application is approved. The modeling software enables OSM to take all the drill-hole information provided by the company, along with the topography of the site, and determine the availability of suitable root zone material from the coal seam to be mined all the way to the surface. For every parameter listed in the permit application – pH, selenium, and so on – it is determined whether there is enough suitable root zone material and if it is available in a timely manner during the mining and reclamation process. The modeling allows OSM to tell the good material from the bad, and gives insight for how the material is to be managed. The modeling program also identifies the location of the suitable material. A calculation is made of exactly how much is needed, where it goes, and whether the operator's material handling program is adequate. Approval is granted based on that determination, and the coal companies ensure that an adequate amount of the "good stuff" ends up on top (see Figure 15.5).

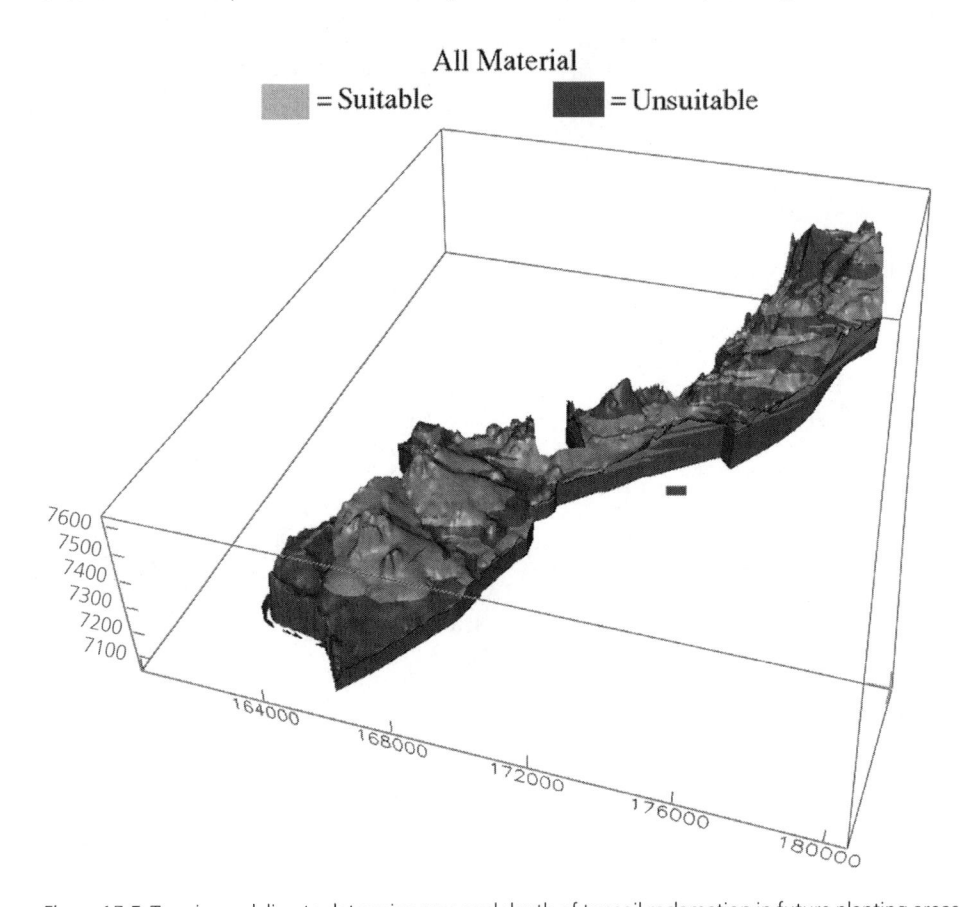

Figure 15.5 Terrain modeling to determine area and depth of topsoil reclamation in future planting areas.

Note

1 Surface Mining Reclamation and Control Act of 1977, Public Law 95-87.

Part IV
Future directions and programs in US reclamation policy and law

Chapter 16
The land revitalization initiative:
landscape design and reuse planning in mine reclamation

Edward H. Chu

Introduction

Over the last three decades, the Environmental Protection Agency (EPA) has expanded its focus from the goal of cleaning up contaminated lands, including hazardous mining sites, to incorporating reuse planning in the cleanup process. The EPA's Land Revitalization Initiative was formed to promote revitalization goals across the Agency's many cleanup programs and to remove or minimize barriers to site reuse. The EPA's Office of Superfund Remediation and Technology Innovation (OSRTI), Abandoned Mine Lands (AML) Team, and Office of Brownfields Cleanup and Redevelopment (OBCR) have been successful in working with partners to revitalize former mining properties in ways that restore ecosystems, preserve history, and spur economic development. With these partners, the EPA is developing strategies to overcome mine reuse challenges and exploring innovative reuse approaches.

Mining lands in the United States

Although no definitive inventory of the number of hard-rock mines on federal lands is available, a 1996 U.S. Geological Survey analysis estimates a total inventory of approximately 209,000 US mines, 48,000 of which are considered past producing hard-rock mines (U.S. Geological Survey 1996). In addition, an estimated 1.5 million acres of disturbed land at abandoned coal mines are spread across the country (U.S. Department of the Interior 2000: 7). Associated with these mining sites and their acid-mine-drainage issues are hundreds of miles of "dead" streams, in which few organisms exist. Environmental contamination, as well as physical hazards, such as open mine shafts and abandoned structures, is typical at former mining sites. Acid mine drainage can affect watersheds and streams, degrade ecosystems, and threaten public water supplies. Individual cleanup solutions can be complex and expensive in accordance with each site's unique flow rates, contaminants, and topographic and climatic conditions. In addition, most mines are located in rural communities, where cleanup and reuse can be further complicated by limited investment, insufficient infrastructure, and complex land-ownership issues.

Evolution of the regulatory framework to incorporate reuse

The legal framework supporting the reclamation of former mines is complicated by the fact that there are a number of regulations governing the various components of the reclamation and reuse process, and each has its own mandates and operating principles. With the exception of the Surface Mining Control and Reclamation Act (SMCRA), all of the key regulations involved in mine cleanup and reuse have a broad scope; though their influence covers mine lands, they were not created solely to support actions at mines.

Historically, the primary regulations that have impacted mine cleanup are the Clean Water Act (CWA), SMCRA, and the Comprehensive Environmental Response, Compensation, and Liability Act (CERCLA). The CWA regulates the discharge of pollutants into navigable waters, including discharges associated with mines. SMCRA ensures that active coal mines comply with reclamation standards during closure activities. It also established a tax on active mines, creating the Abandoned Mine Land (AML) Fund, which is used to fund reclamation associated with pre-1977 coal-mining activities. A portion of the fund is directed to states for the reclamation management of their priority coal sites. States certifying that all coal reclamation has been completed may then use their AML monies for non-coal reclamation. CERCLA, more commonly known as Superfund, provides funding for mine cleanups either through payment for or implementation of cleanups by responsible parties. It also gives natural-resource trustees the ability to pursue those responsible for specific, documented damages, and it enables federal land managers to clean up former mines on federal lands. Superfund has succeeded in cleaning up many of the nation's most contaminated sites. A series of court cases in the early 1990s, however, has resulted in less certainty among lenders and developers about the potential reach of CERCLA liability, leading to a growing reluctance to redevelop properties where low levels of contamination are present or even perceived to be present (LaRosa et al. 2006: 59). The term "brownfields" was coined at about this time to describe former commercial and industrial properties that remain idle as a result of liability concerns.

In 2002, the Small Business Liability Relief and Brownfields Revitalization Act was signed into law, providing a framework of liability clarification, incentives, and financial assistance to encourage the cleanup and reuse of contaminated lands. Now referred to as the "Brownfields Law," it expands the definition of *brownfields* to include mine-scarred lands, providing a new legal and financial tool for the cleanup and revitalization of mining properties and communities. The enactment of the Brownfields Law highlighted the shift in public approaches to addressing contaminated lands in the United States from environmental cleanup as the single goal to an increasingly institutionalized emphasis on cleanup and reuse. Further promoting this shift is the Land Revitalization Initiative, which the EPA created in 2003 to help restore and return contaminated – and potentially contaminated – properties to America's communities for beneficial and productive purposes (U.S. Environmental Protection Agency 2006a). Together, the Brownfields Law and the Land Revitalization Initiative encourage environmentally protective reuse of sites, with the farther-reaching goal of benefiting the affected local communities.

Action agenda for the EPA's Land Revitalization Initiative

The purpose of the EPA's Land Revitalization Initiative is to coordinate and promote the efficient cleanup and reuse of contaminated properties across the EPA's cleanup programs and to remove or minimize the barriers to site reuse. In order to realize the goals of the Initiative, Land Revitalization Staff Office (LRS) has identified the following ten actions on the part of the EPA:

1 Promoting land revitalization as a national policy by ensuring that reuse options are considered in the evaluation of site-cleanup alternatives.
2 Committing the necessary resources to address reuse as a top priority in cleanup decisions. In addition to Brownfields grants, LRS will try to develop incentives for private investment and explore ways to better leverage other federal funding.
3 Developing comprehensive policies and programs to address unintended cross-jurisdiction and cross-program barriers to the safe reuse of previously contaminated properties.
4 Promoting the safe, long-term reuse and stewardship of sites by establishing and main-

taining appropriate engineering and institutional controls, thereby protecting future generations from inappropriate site reuses.
5 Promoting sustainable reuse to prevent recontamination and to minimize other environmental problems that can result from certain reuses. Some examples of sustainable reuse include energy-efficient buildings, smart-growth community development, wildlife habitat, and recreational green space managed in environmentally sound ways.
6 Developing and promoting a land revitalization research agenda that improves stakeholders' understanding of and ability to reuse contaminated or potentially contaminated sites.
7 Building partnerships to leverage knowledge, expertise, and resources in the revitalization of sites. Such partnerships include government-to-government partnerships at the local, state, tribal, and federal levels, as well as government partnerships with non-government, private-sector, and community organizations.
8 Assisting the identification of contamination, cleanup activities, reuse options, and long-term stewardship by expanding community capabilities through improved public-involvement tools and information systems. The EPA will work with states and tribes to develop a Web-based tool that provides communities, investors, and developers with a national inventory of sites with reuse potential.
9 Expanding and promoting educational and training programs that provide insight into land-revitalization strategies, addressing topics such as the development of environmentally challenged properties, risk-management tools, and financing possibilities.
10 Promoting attempts to measure and report the status and impacts of efforts to revitalize properties. The EPA is working to obtain a more accurate picture of the positive impacts of returning idle land to productivity by generating cleanup statistics, examining partnerships and sources of funding, and quantifying benefits to surrounding communities including cleanup statistics, partnerships and sources of funding, and benefits to surrounding communities.

Founded on the experience and lessons learned in the implementation of the Agency's many cleanup programs, the Land Revitalization Initiative provides a way to share models and approaches across the Agency's various efforts. Mine-revitalization projects support the goals of the Initiative by encouraging private investments and promoting the safe, long-term, sustainable reuse of contaminated sites. The EPA's OSRTI, AML Team, and OBCR, in particular, demonstrate the Agency's ability to work closely with mining communities to integrate reuse strategies and landscape design into revitalization activities that support the Initiative's goals while realizing significant environmental, economic, and community benefits.

Benefits of reuse planning and landscape design to mine revitalization

Significant benefits can be realized in the revitalization of former mining sites, especially when reuse and landscape design are incorporated into the cleanup and redevelopment process. With proper landscape design, revitalization can improve the likelihood of ecological success by reducing erosion and preventing contaminants from migrating. Landscape design can also be used to maintain a mining site's historical and cultural legacies. Hiking or walking trails, for example, that navigate visitors past mining artifacts may help maintain a site's integrity while enabling visitors to experience a part of history. In addition, integrating landscape design into the early stages of land-revitalization efforts ensures the efficient planning of economic-development projects; optimal reuse options are determined through an understanding of the physical opportunities and constraints of a property.

The following case studies demonstrate the EPA's encouragement of linkages between cleanup and reuse, the type of support the EPA has provided and will continue to provide in the revitalization of former mining properties, and the importance of landscape design in meeting these objectives.

Mining revitalization and the restoration of ecosystems

Former mines present an opportunity to restore ecosystems, specifically when the properties and parcels to be revitalized are within, between, or adjacent to areas already demarcated as conservation lands or natural habitats. Former mining lands that are not located near such areas can present new opportunities for the preservation of valuable habitat or for the protection of sensitive species. The Silver Bow Creek project in Montana, and the Copper Basin Tailings and Ocoee River project in Tennessee, exemplify successful efforts at ecological restoration.

Habitat restoration, Silver Bow Creek, Montana

For more than a hundred years, Silver Bow Creek was used as a channel for mining and smelting and for industrial and municipal wastes. The Creek's waterways carried approximately 20 million tons of tailings and other mining wastes into the headwaters of the Clark Fork River. In 1983, after a failed attempt to slow the harmful effects of the tailings on the river, the site was placed on the National Priorities List (NPL), which is the EPA's list of national hazardous-waste-cleanup priorities. In all, the Silver Bow Creek Superfund site encompassed twenty-six miles of stream and streamside habitat (U.S. Environmental Protection Agency 2000b).

A strong partnership between the EPA OSRTI and the Atlantic Richfield Company (ARCO) led to the cleanup and ecological restoration of the Silver Bow Creek site. ARCO agreed to pay $1.7 million toward the creation of 400 acres of new wetlands, which today provide an important habitat and breeding area for songbirds and osprey (see Figures 16.1 and 16.2). The site has also become home to 230 types of resident and migratory wildlife. The area has new bike paths, a self-guided walking tour, and other recreational components that integrate people into the site's restored ecology.

Copper Basin Tailings and the Ocoee River, Tennessee

The Copper Basin Tailings site in Tennessee was a fifty-square-mile tract of land that became devoid of biological life after years of mining. The site included the Ocoee River system, which was contaminated by two creeks that were draining the Copper Basin area. Both of these creeks, the North Potato Creek and the Davis Mill Creek, were so acidic that a biological survey found only one living organism between them (U.S. Environmental Protection Agency 2005a: 1). Throughout the entire project, the EPA OSRTI's partners, including representatives from the community, kept in mind the end goal: to allow future reuse of the land and waters. In order to create a habitat suitable for wildlife, native grasses and trees were planted on a vast area of previously unvegetated mine tailings (see Figure 16.3). Additionally, source-control efforts at

Figure 16.1 Silver Bow Creek, Montana. Construction of new wetlands.

Figure 16.2 Silver Bow Creek, Montana. Completed wetlands construction project.

Figure 16.3 Copper Basin Tailings and the Ocoee River, Tennessee. Tree planting for wildlife habitat.

affected tributaries, as well as a wetland-treatment system for a tributary to North Potato Creek, have improved the water quality of the Ocoee River and allowed a whitewater tourist industry to thrive. On land that was once devoid of vegetation and clear waters, trees have taken root and aquatic life is gradually returning.

The ongoing investigations and cleanup efforts at the site are made possible by a Memorandum of Understanding and related legal agreements between the EPA, the Tennessee Department of Environment and Conservation, and OXY Oil and Gas USA, Inc., a corporate successor to Cities Services Company, one of the site's former operators.

Mining revitalization and the preservation of history

Some of today's towns – and even cities – were established around startup mining operations, and restoring the evidence of an early mining industry enables communities to preserve their past. The Anaconda Smelter site, in Montana, and the California Gulch site, in Colorado, are examples of historical-reuse projects that were made possible through the support of the EPA OSRTIs.

Anaconda Smelter, Montana

Established in 1884, the Anaconda Smelter once employed thousands of people, serving as the backbone of the local economy in Anaconda, Montana. After smelting operations ended in early 1980, the site was abandoned with more than 1.4 million cubic yards of soil, slag, and flue dust, contaminated with heavy metals such as arsenic, cadmium, copper, lead, and zinc (U.S. Environmental Protection Agency 2000a).

In order to redevelop a portion of the site, ARCO enlisted the help of golf champion Jack Nicklaus. Taking advantage of the area's beautiful mountain vistas and unique historic characteristics, Nicklaus designed a golf course that preserves and highlights the area's smelting history. In doing so, he filled the bunkers with more than 14,000 cubic yards of inert smelting slag, ground down to the texture of sand (see Figure 16.4). A hiking trail that describes Anaconda's smelting heritage and historic copper-mining operations also winds around the course. This innovative use of a former mine site attracted numerous developers to the area and encouraged local business owners to make improvements to their properties. In developing the cleanup plan and orchestrating an agreement to address liability concerns the EPA OSRTI worked closely with mine-property owners, the community, and county officials.

California Gulch, Colorado

As early as 1857, the historic mining town of Leadville, Colorado, was mined extensively for lead, gold, silver, copper, zinc, and manganese. After the mining operations ended, high levels of lead and other mining wastes in the soils posed risks in residential and commercial areas, while acid mine drainage contaminated the nearby Arkansas River, destroying native vegetation and wildlife habitat and threatening the water supplies for recreation, livestock, irrigation, and public drinking. In 1983, the EPA added the California Gulch site to the NPL.

Initially, the Leadville community opposed the site's listing on the NPL. Many of those concerns dissipated, however, once ASARCO, one of the site's potentially responsible parties, began to work with the community and the EPA OSRTI to find remedies that would not only protect human health and the environment, but would also accommodate community goals for the area. The site's restoration involved partnerships among stakeholders, city and county authorities, area landowners, local civic and historic organizations, and state and federal agencies. In 1998, to facilitate a reuse in the community's best interest, the EPA and the state signed two agreements that ensured public access to the site's restored open space.

Today, a unique twelve-mile trail – the Mineral Belt Trail – loops around mining artifacts and

Figure 16.4 Anaconda Smelter, Montana. Completed golf course.

historic mine-tailings piles within the California Gulch site. Its layout links historic landscapes, buildings, structures, artifacts, and individual locations that distinguish Leadville as Colorado's premier mining camp. During the winter, the path is groomed and enjoyed by recreational skiers (see Figure 16.5).

Mining revitalization to spur economic development

When a mine closes, its surrounding community is forced to transition away from a mining-based economy and to explore other investment and job-creation opportunities. Because former mine sites sometimes have an established infrastructure and are located on large, flat tracts of land, often in otherwise rugged landscapes, they can be an asset to economic-development planning. Although there are few examples of mining sites that have been reused for traditional economic-development purposes, such projects are underway. Currently the EPA is supporting both a multi-reuse development in Hazleton, Pennsylvania, and a renewable-energy facility in Nevada.

The Cranberry Creek Gateway Park, Hazleton, Pennsylvania

The Cranberry Creek Gateway Park is a 366-acre, abandoned anthracite mine site located adjacent to the primary entrance of the City of Hazleton, Pennsylvania, a community whose economy was based on coal mining and is now in the midst of an economic transition. The vision for redevelopment consists of a mixed-use complex with residential and commercial components. It is hoped that, by eliminating blight at the community's gateway, investment in the

Figure 16.5 California Gulch, Colorado.

area will increase. Currently underway are a market feasibility study and a review of physical land attributes, which aim to characterize real-estate and economic-market conditions, to identify feasible development options, and to determine where specific components should be located on the property. There may be, for example, parcels of land that would require land compaction to make them stable enough to hold a building; such areas will likely be prioritized for recreational reuse (see Figure 16.6).

The EPA OBCR and AML Team are supporting this project through the Federal Brownfields Partnership Mine-Scarred Lands Initiative, a collaboration of federal agencies on six demonstration projects across the United States (U.S. Environmental Protection Agency 2005b: 1). This partnership developed a technical and financial plan for the Cranberry Creek site, which outlines potential resources from federal, state, and local agencies and describes how they can be efficiently combined.

Renewable Energy Park, Beatty, Nevada

The Bullfrog Mine has a 100-year mining history in Beatty, Nevada, a small town of approximately 1,200 residents, located 110 miles northwest of Las Vegas. After the mine's closure, the mining company transferred eighty-one acres of cleaned-up mine lands to the community (see Figure 16.7). The community is now focusing on reusing the site for a renewable-energy facility, because the area has significant solar, wind, and geothermal resources. It is hoped that the existing transmission lines and buildings will support some of the infrastructural needs of the renewable-energy facility. Such a project would offer new and unique economic opportunities to the community. Partners in the project are considering an expansion of the project that would develop a regional renewable-energy corridor, stretching across many communities and providing a significant energy resource for Nevada and California.

As with the Cranberry Creek Gateway project, the EPA OBCR and AML Team are supporting the revitalization effort in Beatty through the Federal Brownfields Partnership Mine-Scarred Lands Initiative. The Initiative has assisted this project by engaging and coordinating wide-ranging partnerships with the mining company, the Nevada Energy Office, the U.S. Department of Energy, the Bureau of Land Management, and national energy laboratories and utility companies. The EPA also aided the development of an action plan for the site, which outlines the next

The land revitalization initiative

Figure 16.6 Cranberry Creek Gateway Park, Hazleton, Pennsylvania. Former coal refuse and railroad; future site of mixed-use development project.

Figure 16.7 Renewable Energy Park, Beatty, Nevada. Potential area for renewable energy facility.

steps for stakeholder engagement, funding sources, technical feasibility, and general project development.

EPA initiatives that further mine revitalization

The EPA supports the reuse of former mine properties through initiatives that extend beyond the scope of the Agency cleanup programs. The EPA's Good Samaritan Initiative, for example, is developing innovative tools that better enable volunteers at mine cleanups, while the Federal Mining Dialog and Federal Brownfields Partnership Mine-Scarred Lands Initiative are promoting collaboration among federal agencies involved with mine revitalization. Lastly, the EPA is focused on developing cleanup-program performance measures to evaluate revitalization progress more easily on former mining sites.

Collaboration among federal agencies: the EPA AML Team, Federal Mining Dialog, and Federal Brownfields Partnership Mine-Scarred Lands Initiative

The EPA's AML Team provides EPA Headquarters and ten regional offices access to expertise on issues at abandoned mine sites, with the goal of setting priorities for the evaluation, cleanup, and redevelopment of the sites. The Team serves as a focal point for coordinating and facilitating EPA policy, funding, process, and technical issues among a project's stakeholders; these stakeholders comprise, but are not limited to, states, tribes, the National Mining Association, the Mineral Policy Center, the Bureau of Land Management, the U.S. Department of Agriculture's Forest Service, and the Western Governors Association.

The AML Team helped to establish a formal Federal Mining Dialog in the mid-1990s to foster cooperation and coordination within the federal sector, especially where former mine lands involve more than one federal agency. The partners of the Federal Mining Dialog include the Bureau of Land Management, the EPA, the U.S. Department of Agriculture's Forest Service, the National Park Service, the Department of Justice, and other federal agencies and departments. The Dialog meets on a quarterly basis to communicate about issues related to former mine lands located within federal authority, and it occasionally holds larger conferences for federal partners.

The Federal Brownfields Partnership Mine-Scarred Lands Initiative is a similar collaboration among federal partners. The establishment of this interagency partnership in 2003 was closely tied to the signing of the Brownfields Law, which provided new legal and financial tools for the cleanup and revitalization of mining properties. The Mine-Scarred Lands Initiative is currently involved in six demonstration projects across the country in order to better understand the challenges that mining communities face in the revitalization process and to build more effective collaboration among the federal agencies. The six agencies participating in the Initiative are the Appalachian Regional Commission, the U.S. Army Corps of Engineers, the U.S. Department of Agriculture, the U.S. Department of Housing and Urban Development, the U.S. Department of the Interior, and the EPA. Individual projects have involved other federal agencies, along with a wide range of state, local, and private-sector partners. The EPA's AML Team and OBCR, for example, led the development of the online Mine-Scarred Lands Initiative Tool Kit, which shares lessons learned in this collaborative effort (U.S. Environmental Protection Agency 2006b).

Measuring revitalization success with the EPA's performance measures

As part of its goal to return formerly contaminated sites to long-term, productive use, the EPA is implementing new land-revitalization performance measures and indicators. To date, thousands of acres of land, including hazardous mining sites, have been assessed or cleaned and deemed appropriate for present and future uses. The current process for measuring and reporting revitalization accomplishments varies by cleanup program; the new performance measures will establish a uniform reporting mechanism that more accurately details successes while promoting future efforts.

Innovative opportunities to support mine revitalization

The EPA recognizes that public funding sources for mine cleanup and reclamation are limited. Therefore, to fully capitalize on the economic-development and environmental opportunities associated with mining sites, innovative reuse strategies that maximize economic gain may be considered. Selecting reuse strategies that yield significant economic value may encourage lenders and investors to finance cleanup activities based on the expected return on investment. Some emerging strategies for the enhancement of mine revitalization include renewable-energy development, carbon sequestration, and wetland banking. The EPA is monitoring the evolution of these innovative approaches and exploring the coordination, partnership, and other support efforts that further the revitalization of the land occupied by former mines.

Renewable-energy facilities

According to the U.S. Department of Energy's *Annual Energy Outlook*, domestic use of electricity is expected to increase by more than 40 percent by 2025, likely requiring the construction of new power-generation facilities on thousands of acres of land (U.S. Department of Energy 2006). Many of these new energy facilities could potentially be developed on former mine lands. There are several reasons former mine sites are well-suited for reuse as renewable-energy facilities: the transmission capacity and infrastructure are often already in place at mine sites; renewable energy is an economically viable reuse for mining sites that typically have significant cleanup costs or low real-estate-development demand and renewable-energy facilities spur investment and create jobs.

Near Wilkes-Barre, Pennsylvania, the Bear Creek Wind Farm was constructed on former coal-mining lands. Operational since February 2006, the wind farm features twelve wind turbines that will provide more than 70 million kilowatts of clean, renewable, and domestically produced electricity for more than 8,500 homes each year (New Wind Energy 2006; Pennsylvania Wind Working Group 2006).

Carbon sequestration

Receiving "carbon credits" for reforestation and carbon sequestration can also provide incentives for energy investors or other utilities to fund mine-reclamation projects. Companies can buy or generate these credits, which are then sold or traded to offset carbon-dioxide (CO_2) emissions as needed. For example, in 2001, Allegheny Energy initiated a reforestation and sequestration pilot in western Pennsylvania, whereby fifteen acres were replanted with more than seven thousand pine and spruce trees and two acres were planted with warm-season grasses. The Pennsylvania Department of Environmental Protection calculated that, at maturity, the trees could remove sixty-four tons of CO_2 per year. These carbon credits are now registered with the state voluntary greenhouse-gas registry. Beyond the financial benefits of reforestation in reclamation activities are the environmental benefits achieved through the absorption of carbon from the soil into the trees, the enhancement of wildlife habitat, and the improvement of air and water quality (U.S. Environmental Protection Agency 2004a: 1).

Wetland banking

Wetland banking is the restoration, creation, or enhancement of wetlands to offset future development impacts on other wetlands. The strategy was designed as a compensatory mitigation mechanism to support wetlands preservation. This innovative option for the redevelopment of former mine lands can provide financial rewards through the sale of bank "credits" to landowners or developers who need to compensate for damage done to wetlands in prior development projects. Mine owners can use the revenue from these credit sales to help fund site cleanup. Transforming former mines into wetlands can also create habitats for plants and animals and help to remove harmful metals from contaminated waters (U.S. Environmental Protection Agency 2004b: 1). Although there are, at present, few examples of wetland banking on former mine sites, there appears to be significant potential for the strategy, since many former mines have been successfully converted into wetlands. The previously described Silver Bow Creek project, for example, created 400 acres of new wetlands in an area of Montana that once carried approximately 19 million tons of tailings and other mining wastes into the headwaters of the Clark Fork River (U.S. Environmental Protection Agency 2000b).

Conclusion

Public approaches to mine-land revitalization have shifted from focusing on the mere cleanup of contaminated lands to ensuring their cleanup as well as their reuse for beneficial and productive purposes. This transition has enabled many communities to realize their revitalization goals. The EPA has provided critical support to these mining communities in their efforts to revitalize the land by restoring ecosystems, preserving local history, and spurring economic development. The EPA has expanded this support through a range of collaborative initiatives and will continue to explore innovative strategies for supporting abandoned-mine revitalization.

Bibliography

LaRosa, C., Arial, J., and Larsen, K. (2006) 'Integrating community capacity building and reuse planning into the reclamation of American mine lands: how US regulators are cleaning up with communities,' *Mining Engineering* 58, 3: 59–64.

New Wind Energy (2006) 'The Bear Creek Wind Power Project.' Online. www.newwindenergy.com/windfarm_bearcreek/index.html (accessed September 6, 2006).

Pennsylvania Wind Working Group (2006) 'Wind Farms in Pennsylvania.' Online. www.pawindenergynow.org/pa/farms.html (accessed September 6, 2006).

U.S. Department of Energy, Energy Information Administration (2006) 'Annual Energy Outlook 2005, Market Trends – Electricity Demand and Supply.' Online. www.eia.doe.gov/oiaf/archive/aeo05/electricity.html (accessed July 18, 2006).

U.S. Department of the Interior (2000) 'Office of Surface Mining Strategic Plan: 2000–2005.' Online. www.osmre.gov/pdf/strategicplan00.pdf (accessed July 18, 2006).

U.S. Environmental Protection Agency (2000a) 'Anaconda Smelter One-Page Summary.' Online. www.epa.gov/superfund/programs/recycle/success/1-pagers/anaconda.htm (accessed July 18, 2006).

—— (2000b) 'Silver Bow Creek/Warm Springs Ponds One-Page Summary.' Online. www.epa.gov/superfund/programs/recycle/success/1-pagers/bowcrk.htm (accessed July 18, 2006).

—— (2004a) 'Carbon Sequestration: A Local Solution with Global Impacts.' Online. www.epa.gov/superfund/programs/aml/revital/cseqfact.pdf (accessed July 18, 2006).

—— (2004b) 'Wetland Banking at Former Mine Lands: An Ecological Solution with Economic Benefits.' Online. www.epa.gov/superfund/programs/aml/revital/wlfact.pdf (accessed July 18, 2006).

—— (2005a) 'Copper Basin Mining District Case Study.' Online. www.epa.gov/superfund/programs/aml/tech/copperbasin.pdf (accessed July 18, 2006).

—— (2005b) 'Mine-Scarred Lands Revitalizations: Models Through Partnerships.' Online. www.epa.gov/brownfields/policy/initiatives_sb.htm (accessed July 18, 2006).

—— (2006a) 'Land Revitalization Initiative.' Online. www.epa.gov/landrevitalization (accessed July 18, 2006).

—— (2006b) 'Mine-Scarred Lands Initiative Tool Kit.' Online. www.epa.gov/aml/revital/msl/index.htm (accessed December 7, 2006).

U.S. Geological Survey (1996) 'National Overview of Abandoned Mine Land Sites Utilizing the Minerals Availability System (MAS) and Geographic Information System (GIS) Technology.' Online. http://pubs.usgs.gov/of/1996/ofr-96-0549/ofr-96-549.html#anchor648696 (accessed August 30, 2006).

Chapter 17
The legal landscape

Robert W. Micsak

From my earliest involvement with landscape architecture, I have been incredibly interested in the interplay between the law and the landscape. In few places is the impact of the law more apparent than in the mining and reclamation business. Millions of acres of United States land have been affected by mining in the last hundred years, and millions more will be affected in the next century. There is no single business operation today that approaches the magnitude of landscape change that can occur at a modern mine. Mountains can literally be turned into valleys and valleys into mountains. A single mining operation can move across thousands of acres, completely removing and reconstructing the surface. Nowhere else provides a greater opportunity for the landscape architect to consider the future landform and its use on such a scale. For the practitioner to be truly effective, however, design talent alone is not enough. A thorough understanding of the complex of laws and regulations that mandate the myriad of legislated environmental, social, and economic objectives is required.

In the mineral-development business, there is a tremendous array of laws that affect virtually every aspect of the enterprise, from worker safety and initial exploration, through operations, to final reclamation and mine closure. This chapter will briefly explore only those laws and regulatory programs that most directly impact land-use planning and reclamation of metal and coal mines in the United States.[1] The two primary types of mine that will be detailed here are surface coal and metal (often referred to as "hardrock").

In order to set the stage for what can and cannot be done in the reclamation process, it is necessary to describe some of the key characteristics of these types of mining operations, as well as the landscape implications that arise from them. The more significant laws that drive the mine-siting, relevant operations, and configuration decisions regarding any particular mining proposal will then be addressed. Finally, some of the tools available to the mineral developer to gain greater control over the regulatory process and mine-planning and -closure activities will be highlighted.[2]

When the landscape architect is considering mine reclamation and planning, a few factors ought to be kept in mind. First and foremost, the mine planner must extract the minerals where they find them. Sensitive areas, rough topography, wetlands, proximity to infrastructure, proximity to communities, involvement of endangered species, the moisture of the environment, ownership of the land, contaminants in the rock or mineral-bearing strata – the good, the bad, and the ugly – all have to be dealt with as they are delivered up. Beyond the mineral location, however, a broad range of options is available to deal with the variety of issues.

In the case of modern mines, there are large tonnages of soil and rock that must be moved: typically hundreds of millions of tons of material, but in some cases billions of tons of material at a single operation. A single modern mining-haul truck is today capable of carrying on the order of 350 tons of material in its bed (enough to fit a good-size family home). Therefore two considerations ought to govern any planning decisions:

1 The distance the material is hauled factors significantly in the cost of any operation. Shorter haul distances reduce costs. Therefore, haul distances should be minimized in any reclamation plan.
2 Every haul costs lots of money. Large tonnages means that re-handling material just one time can add hundreds of millions of dollars to the cost of the operation. Even though some modern mining trucks can haul more than 300 tons of material in a single load, mine operators do not want to "re-handle" material. In other words, once it is picked up and put in place, it ought to stay there.

Coal mining

Coal mines can be underground or surface operations. The focus here is on surface mines, because they generally impact a greater acreage of surface lands than any other type of mine and therefore offer the best opportunities for landscape planning. Single, operating surface coal mines can range in size from a few acres to several thousand acres. Coal that is surface-mined is usually laid down in layers, or deposits, of varying thickness.[3] The typical operation involves stripping away the topsoil and storing it for later replacement and reclamation. In some instances, particularly in US Midwestern and Eastern mines, where soils are deeper and more fully developed, there may be more than one soil horizon,[4] each of which requires separate removal and storage. After the topsoil is removed, the overburden, or the non-coal-bearing materials that overlay the first economically viable seam of coal, is removed. At larger mines, this overburden removal is often conducted using draglines.[5] This last operation exposes the coal, which is then removed using large bulldozers, shovels, or front-end loaders, which make a clean separation of the coal down to the next layer of non-coal-bearing material. If there are seams of coal below the top seam, then the interburden is removed, thereby exposing the deeper coal seam.

The Surface Mining Control and Reclamation Act of 1977

The Surface Mining Control and Reclamation Act (SMCRA) of 1977 (30 USC §§ 1201 *et seq.*) is the principal law governing reclamation of the surface effects of coal mining (including underground operations) on lands in the United States. It applies to surface impacts on all lands, both public and private.[6] It requires, *inter alia*, that the land be restored to its "approximate original contour" (AOC) (30 USC § 1265b (3)).[7] Thus, as mining operations advance across a landscape, the reclamation process begins with the replacement of the inter- and overburden, accompanied by rough grading to the AOC of the pre-mining condition. The topsoil horizon is then laid down on top of the overburden, and the final planting and seeding is undertaken.

One of the most significant provisions of the SMCRA is it allows for an exemption, if certain conditions are met, from the AOC requirement. Briefly, Section 515(e) of the SMCRA statute declares that there is available a

> variance from the requirement to restore [lands] to approximate original contour . . . for surface mining of coal where the owner of the surface knowingly requests, in writing, as

part of the permit application, that such a variance be granted so as to render the land, after reclamation, suitable for an industrial, commercial, residential, or public use (including recreational facilities).[8]

This variance offers tremendous, really incredible, opportunities for the forward-thinking mine planner, as well as the landscape architect. It is a tabula rasa for post-mine reuse and landscape reclamation creativity. The entire landform is created and re-created in the reclamation process. Golf courses, recreational facilities, outdoor theaters, and a host of other projects have and can result from well-thought-out reclamation and mine-closure planning. Creative, cost-effective plans can procure community, mining-company, and NGO support.

Coal mining on federal and state lands

SMCRA and its implementing regulations govern the permit process for approval of surface-coal-mine plans on all federal and private lands. OSM is responsible for the permitting of surface coal mines on federal lands but must consult with the appropriate federal land-management agency in granting permit approval.[9]

SMCRA provides for states to take over the regulatory program from OSM, though OSM oversight continues after the delegation. To do so, the state must develop a program – including the adoption of laws and regulations that are no less stringent than those of the federal program – provide adequate funding for the program, and gain OSM's approval.

Metals mines

It is not unusual for metals mines to operate for decades, and some of the larger mines have been in operation for a hundred years or more (for example, the Colorado CC&V and Kennecott's Bingham pit in Utah, the pit of which can be seen from the moon). There is a gold mine in the Transylvania district of Romania that has operated off and on for nearly two thousand years. Their long life span raises reclamation and closure issues that are different from those discussed with respect to coal-mining operations. The most significant of the issues that arise from the length of operation is the inability to fill the metal mine pits back in. Just think about it: if it takes a hundred years to dig the ore out, it ought to take about the same amount of time to fill it back in, assuming the use of similarly sized equipment. As a consequence, it is, in most instances, simply unrealistic, not to mention cost-prohibitive, to backfill the pits of hardrock mines. When there are several smaller pits and they can be mined in sequence, however, there is an opportunity to backfill.[10]

Metals mines are characterized by the processing of rock that contains economically recoverable amounts of a given metal, which is called ore. It is the extractability of the metal at a profit that makes it ore.[11] Ore is stacked and processed on a lined pad by leaching solution through the ore, which dissolves the metal; the resulting solution is then transported to a recovery, or processing, facility. Alternatively, the actual ore can be passed directly through a processing facility. There it is ground, or milled, into finer particles and subjected to various forms of chemical or mechanical treatment in order to liberate and recover the metal. The waste material is called "tailing" and is disposed of in a "tailing pond." The processed ore from leaching operations and residual solutions, as well as the tailing, release contaminants that pose important environmentally related closure and reclamation design issues for the miner.

In the cases of process-facility and leach-pad-facility closure, the focus is upon neutralization or containment during the final closure and reclamation phase. In practice, such facilities contain residual amounts of the metals not recovered by the mining process; no process removes 100 percent of the minerals from the ore. These facilities also contain residual amounts of process solutions, which include the chemicals used to dissolve the metals (cyanide is a typical component). They may also contain chemicals that stabilize the solutions as they move through various temperature regimes or offset the adverse impacts of the mineral-recovery process or of other minerals in the ore. Typically, the pH of the process solutions is outside the regulatory limits for direct discharge into surface watercourses. Therefore, these facilities are typically lined and equipped with leak detection and solution-recovery and -control systems. They require very specialized closure planning and design, in order to assure that all contaminants are neutralized and that no contaminants escape after closure. The goal is to eliminate any future risk of contamination to air, water, or soil. These facilities also offer limited opportunities for the landscape architect to develop surface facilities and land uses that do not pose any risk of subsequent contamination of the surrounding lands or underlying or adjacent watercourses. If the contaminant material can be completely neutralized, then there is great opportunity to re-grade and reclaim the area in question in any way that is consistent with the post-mining land-use plan.

The rock that does not contain any economically viable mineralization is called, not surprisingly, "waste rock." Waste rock is separated from ore through an elaborate sampling and assay program; then it is usually deposited in its final resting place, where it awaits final grading and reclamation. Thus, planning where and when to deposit waste rock is a very important decision for the mine planner, since no miner wants to place waste rock over an area that might contain ore beneath it – or which might otherwise require the rock to be re-handled (remember rule 2 above!).

Waste rock, too, may contain mineralization that can, when water percolates down through it, carry metals or other contaminants to surrounding lands and watercourses. Many rivers and streams have been polluted as a result of this phenomenon. Hundreds of millions of dollars are spent each year to clean up the land and water contaminated by this type of drainage from old, abandoned mining areas, and a lot has been learned about the hazards posed by the interaction between the geochemical and hydrologic regimes over the last hundred years on these sites. The most significant risk, among other things, relates to the sulfide in the waste rock or pits. When exposed to the oxidizing effects of air and water, sulfide can generate acid, which then dissolves the metals and other minerals in the waste rock, polluting receiving streams and other water bodies. Much of the planning of modern metal-mine closure is dedicated to characterizing the rock for its acid-generating potential and then developing mechanisms for eliminating or reducing the pollution risk to acceptable levels. There are several strategies for doing this:

- mixing neutralizing material, such as limestone, with the potentially acid-generating rock;
- creating an impermeable cap on the waste-rock pile, preventing or minimizing water infiltration to levels so low that the oxidation process is stopped or slowed;
- diverting water around the dump, so that little water percolation occurs; or
- creating underdrains, so that water passes under the waste rock and does not infiltrate the potentially problematic material.

Regardless of the methods chosen, waste-rock deposition and configuration are largely unconstrained and offer the landscape architect a broad range of post-mining landform and land-use options.

Hardrock mining on non-federal lands

Hardrock mining on non-federal lands is governed largely by state reclamation laws. A few high-profile environmental disasters at mining operations in the last twenty years have led some states to take steps to strengthen their mining-reclamation and permitting laws. The Summitville disaster in Colorado, in particular, comes to mind, which occasioned the revision of the Colorado Mined Land Reclamation Act (CRS §§ 34-32-101 *et seq.*). The focus of such laws has generally been on identifying environmental risks before the first shovelful of dirt is moved and requiring the mine operator to provide reasonable assurances that those risks are going to be adequately managed and the citizenry and environment protected. Risk identification and management include:

- characterizing waste rock to determine its potential for adversely impacting water quality, in particular its acid-generating capability, and characterizing methods to avoid those adverse impacts;
- assuring that process facilities are designed to eliminate the risk of process-solutions leakage;
- assuring that containment structures can actually contain the appropriate volumes of process solutions and that no spillages or overtopping of containment ponds will occur if pumps are shut down or unusual precipitation events occur;
- providing adequate financial guarantees, so that the full cost of reclaiming the site, including detoxification of all process solutions and facilities, is assured.

Obviously, these requirements exist in addition to those imposed by other laws, such as the Clean Air Act, Clean Water Act, the Resource Conservation and Recovery Act, among others. They serve as an additional layer of protection from the environmental risks that are unique to mining.

One area in which the states have led the way in risk management is that of financial-assurance protections. Bonds, letters of credit, and other forms of financial assurance are used to provide financial protection to the relevant permitting authority in charge of permitting the mine (e.g. Colorado Mined Land Reclamation Board), so that it will be able to finance the mine closure and reclamation if the operator goes bankrupt or is otherwise unable to pay the full cost of this important phase of the operation. Unfortunately, in some historic cases, the agreed-upon amounts were too low to cover the ultimate cleanup costs. Summitville is one of the most significant examples in recent history of the bonding shortfalls that regulatory bodies can face. Most state and federal agencies have now adopted requirements that have significantly increased financial assurances, in the form of financial guarantees from mine operators; the guarantee of a single mine can now exceed 50 million US dollars.

Bureau of Land Management lands and hardrock mines

The Federal Land Policy and Management Act of 1976 (43 USC §§ 1701 et seq.), commonly referred to as FLPMA, applies to lands controlled by the U.S. Department of the Interior's Bureau of Land Management (BLM).[12] At 43 USC §§ 1712 et seq. the statute directs the Secretary of the Interior to develop land-use plans that:

- apply principles of multiple use and sustained yield management;
- give priority to the designation and protection of areas of critical environmental concern;
- develop and use an inventory of the public lands, their resources, and other values;
- consider present and potential uses of the public lands;
- weigh long-term benefits to the public against short-term benefits; and
- provide for compliance with state and federal pollution-control laws.

This directive supports mineral development on BLM lands, including the mining of metal ores, and in general requires the BLM to consider plans submitted for such actions, unless otherwise barred by law.[13]

Surface-management regulations have been adopted to implement this statutory guidance (38 CFR §§ 3809 et seq.). These regulations outline the review and approval process for processing permit requests for metals mines located on BLM lands. They also establish guidelines to assure compliance with the FLPMA prohibition of "unnecessary or undue degradation of public

The legal landscape

lands" (43 USC § 1732(b)). To start the process, an applicant submits a plan of operations, which contains the information required by the regulations, including the life-of-mine plans, waste-rock development and phasing, final pit configuration, environmental-baseline data, impact-mitigation plans, reclamation plans, and post-closure financial assurances. The submission of the operations plan begins the formal mine-permit review, assessment, and approval process. It also triggers the provisions of, *inter alia*, the National Environmental Policy Act, which will be discussed further below.

Forest Service lands and hardrock mines

The Forest Service Organic Act (16 USC §§ 473–78, 479–82, and 551) and the National Forest Management Act of 1976 (16 USC §§ 1600–614) govern, *inter alia*, the regulation of hardrock mines on U.S. Forest Service lands. The Forest Service regulations governing hardrock mining are set out at 36 CFR Part 228 and establish detailed regulatory guidance that is intended to "minimize adverse environmental impacts on national forest surface resources," as mandated by the Organic Act. These regulatory requirements, applicable to submitting and reviewing plans of operations on Forest Service lands, provide extensive information for submittals and public involvement; in the broadest sense, they are similar to the regulatory requirements that are applicable to BLM lands.

National Environmental Policy Act

This law in many ways serves as the great melting pot of all project considerations. The National Environmental Policy Act of 1969, or NEPA (42 USC §§ 4321 *et seq.*), serves to integrate federal land-management decision-making on particular mining proposals with other environmental concerns on the federal, state, and local levels. The key piece of NEPA is set out in Section 102, which states:

> The Congress authorizes and directs that, to the fullest extent possible, all agencies of the Federal Government shall utilize a systematic, interdisciplinary approach which will insure the integrated use of the natural and social sciences and the environmental design arts in planning and in decision-making which may have an impact on man's environment . . . include in every *major Federal action significantly affecting the quality of the human environment, a detailed statement* by the responsible official on –
>
> (i)   the environmental impact of the proposed action,
> (ii) any adverse environmental effects which cannot be avoided should the proposal be implemented,
> (iii) alternatives to the proposed action,
> (iv) the relationship between local short-term uses of man's environment and the maintenance and enhancement of long-term productivity, and
> (v)   any irreversible and irretrievable commitments of resources which would be involved in the proposed action should it be implemented.

The Council on Environmental Quality has adopted regulations implementing its responsibilities under NEPA (40 CFR Parts 1500 *et seq.*), which are binding on all federal agencies.

For operations on federal lands that are expected to have significant environmental impacts, a comprehensive environmental impact statement (EIS) must be prepared. Preparation of the EIS involves procedures for publicly "scoping" the issues and identifying alternatives to be evaluated, and the process results in a Record of Decision (ROD), which determines the content of the plan of operations and operational and mitigation requirements. Care must be taken to assure that all impacts associated with the project are evaluated; otherwise, if the project scope increases or other parameters of the project change, involving impacts that were not analyzed in the prior NEPA documentation, an operator can find itself conducting another NEPA analysis. The preparation of an EIS under NEPA can take anywhere from several months to several years. Proposed operations and activities for which no significant impact is anticipated can be evaluated by the use of an environmental assessment (EA); at times, an EA is used to assist in determining if an EIS should be prepared.

The NEPA process is a very public process, involving a number of public reports, meetings, hearings, and reviews. State, federal, and local considerations are necessarily integrated into each project review and evaluation. A "no action" alternative is always considered, which often raises concern in project proponents that the project will not proceed in any form.

For the landscape architect, the consideration of alternatives under NEPA offers tremendous opportunity. The varying landforms, land uses, and related impacts of each alternative can be considered and evaluated in the context of the social, economic, and environmental objectives of the project.

The Clean Water Act, Section 404

Section 404 of the Clean Water Act (CWA) (33 USC §§ 1251 *et seq.*) deals with the protection of wetlands. Wetlands include the kinds of lands that we all know and recognize as wetlands: lands that contain water, aquatic plant life, ducks, geese, and so on. The regulatory definition extends beyond this, however, to include lands that may in fact be dry yet support aquatic plants or, at the very minimum, only wetland soils. For the land planner, it is important to avoid such areas. Dredging or filling a wetland requires a permit from the U.S. Army Corps of Engineers. There are exemptions for very small or routine wetlands impacts, but every mining operation with which I have been involved has inevitably triggered permitting requirements under Section 404.

The applicant is required to map and characterize the impacted wetland and then to develop a mitigation plan for creating new wetlands, usually at a ratio of two or more new acres for each acre of impacted wetland. Extensive site investigations are then required, in order to assure that the water supply is adequate and that the proper soil type is located or made available at the mitigation site. To meet permit conditions, it is necessary that the created or expanded wetlands survive and prosper for an extended period. Monitoring of plant and animal diversity are also key aspects of meeting permit conditions under Section 404.

Endangered Species Act of 1973

The Endangered Species Act of 1973 (16 USC §§ 1531 *et seq.*) was passed to protect threatened and endangered species. This protection extends from vertebrates to invertebrates and to plants of all sorts, from the noblest creature to the most arcane insects and humblest of grasses. Of critical importance to the land planner is that its protection extends not only to the living creature, but also to the creature's critical habitat, that is, the habitat that is critical to the continued existence and propagation of the species (16 USC § 1532). New species can be added to the list, and recovered species can be removed from it. In many instances, the habitat exists in locations where the species has not been seen for years, if ever.

The Endangered Species Act requires federal agencies to consult with the U.S. Fish and Wildlife Service to ensure that proposed actions do not harm any threatened or endangered species or their critical habitats. If it is determined that a proposed action, such as a mine permit approval, a 404 permit approval, or any other form of US government action, could jeopardize a species or habitat, then the agency and the Wildlife Service must work together to design reasonable and prudent alternatives that address the needs of the species while allowing the project to go forward. Mitigation, habitat acquisition, and funding research into various aspects of the relevant species are but a few examples of the alternatives that have been used to allow a

The legal landscape

project to proceed in spite of potential impacts to the species or its critical habitat. Nevertheless, identifying the types of critical habitat that may exist in an area and understanding that impacts to that habitat are presumptively avoidable can undoubtedly play a significant role for the landscape architect who is involved in planning a mine.

National Historic Preservation Act of 1966

The Advisory Council on Historic Preservation (ACHP) is charged with administering the National Historic Preservation Act of 1966 (16 USC §§ 470 *et seq.*). A key feature of this Act is described in Section 106 and requires federal agencies to review all federal actions that may affect a property listed, or eligible to be listed, on the National Register of Historic Places. This eligibility component and the ACHP consultation process have created significant delays and costs for many mine operators. If historic resources are known or discovered at a site, regardless of whether the site is or is not a listed historic property, then extensive survey, recordation, and preservation steps must be taken. This process allows the significance of the historic resources to be ascertained and a decision to be made regarding the value of preserving that information. However, an agency cannot know if its actions will trigger the conditions of the Act until the survey is complete.

One may note that preservation does not necessarily require in-situ protection but can encompass recovery, photography, drawings, written descriptions, and other forms or methods of preserving historic features or information. These surveys and recordation steps can cost millions of dollars and can take months or years to complete. It is therefore important to identify any known or potentially historic sites, including prehistoric sites, and to take all steps possible to avoid impacting them or else to commence and complete the survey and recordation process in a timely fashion. Phasing the survey and recordation of impacted areas can provide a way for the mineral developer to obtain piecemeal clearance of the project areas while the federal agency still fulfills its obligations under the Act.

Land exchanges and FLPMA Section 106

The authority of the BLM and the Forest Service over land exchanges derives from FLPMA (43 USC §§ 1701 *et seq.*) as amended by the Federal Land Exchange Facilitation Act of 1988. The agencies are given broad discretion to undertake land exchanges where the public interest will be well served. The lands must be within the same state, and they must be of equal value (though values can be equalized by having one party pay up to 25 percent of the land's value to the other). In my experience, the proponent of an exchange cannot expect to receive a payment from the government; and if there is excess value, the developer sometimes makes it into a charitable contribution to the government as part of the exchange. The lands acquired by the government then become public lands, subject to existing laws and regulations.

Conceptually, these are simple exercises. The proponent of an exchange typically consults informally with the federal agency to determine if the agency is interested in disposing of the federal lands that are sought by the proponent. The agency then lets the proponent know the kinds of lands that are of interest to the developer and to the agency itself. The exchange is discretionary on the part of the agency, so acquiring high-value lands is key to a successful exchange. The developer then seeks to acquire the lands or interests in question and proceeds with moving through the regulatory process to consummate the exchange.

From the mineral developer's perspective this law offers several opportunities to:

- gain ownership of land that is key to its operations;
- simplify regulatory requirements by potentially eliminating federal-agency land-management oversight; and

- gain public support for project proposals, since NGOs often become exchange proponents, and thereby project proponents, when there is an opportunity for lands or interests that further their public-interest goals to be moved into public ownership.

In return, the federal government gains lands that are high in public-interest objectives, such as, for example, lands that possess wildlife values, wilderness values, public access, recreation, and open space.

Exchanges are subject to NEPA review, though in many cases prior NEPA documentation related to the mine operations can be relied upon for the purposes of NEPA compliance.

Conservation easements

A conservation easement is a creature of state law. It is a legal agreement, between a landowner and a land trust or government agency, that permanently limits the uses of land in order to protect certain values. It allows the grantor to continue to own and use his or her land and to sell it or pass it on to heirs.

Easements can be used to restrict or limit development, to protect riparian areas, to maintain open space, or to provide for public access or other protections designed to achieve conservation, recreation, or environmental goals. They can be applied to all or any portion of a property.

The restriction of development rights reduces the market value, and therefore the property taxes, of the land. It can be used to preserve agricultural land and lifestyles, especially in places where rising land values cause property values, and related property taxes, to skyrocket. In addition, the grant of the easement can sometimes be deemed a charitable contribution of the fair market value of the easement, thus providing the landowner with the possibility of taking a charitable deduction under state and federal tax law. Some states provide for a tax credit for a portion of the value of the easement.[14]

For the mineral developer and landscape architect, a conservation easement provides a valuable tool for achieving mitigation objectives and support for land exchanges where certain resource values are sought by the government. It costs less than acquiring the entire landholding, and it allows a landowner to keep and operate his or her land. I have seen conservation easements used to protect breeding habitat for endangered fish along riparian areas and to support other elements of land exchanges.

Conclusion

The law clearly plays a critical role in shaping the post-mining landscape. There are a broad range of tools and opportunities available to the landscape architect in planning the reclaimed landscape. The various laws provide considerable flexibility in the creative process, and every landscape architect should have the opportunity to bring his or her design innovations to the world of mine reclamation at some point in their career.

Notes

1 This chapter does not address the impacts that local planning and zoning laws and regulations may have on mining.
2 It should be pointed out that each of the laws and related regulatory programs here mentioned is itself extensive and complex. Together, they encompass hundreds of pages of statutory provisions, thousands of pages of implementing regulations, and thousands of cases that interpret those statutory and regulatory provisions, and this is not even to mention the guidance documents that are equally voluminous. In most instances, legal practitioners spend their working lives involved with only three or four of the regulatory regimes identified here; therefore only the high points of the key provisions of these laws will be addressed.

3 Depending upon the thickness of the deposit, the amount of overburden, the moisture, the ash, and other characteristics of the coal, the mine plan establishes which coal deposits can be mined and sold profitably. If the seams are thin, the overburden too thick, or the quality too low, then it may not be profitable to mine the deposits at a given market price. Evaluating the economics of mining a deposit involves extensive technical analysis before the first shovelful of topsoil is removed. Coal seams range from a few inches to well over a hundred feet in thickness.
4 Soil horizons are determined according to various criteria set out in the applicable federal or state regulatory programs.
5 Draglines are the types of mining equipment that use cables (i.e. lines) to drag a large, flat bucket across the overburden, scraping up the material and casting it back behind the ongoing or advancing mine.
6 SMCRA and its accompanying regulations are administered by the Office of Surface Mining (OSM) of the U.S. Department of the Interior. Each state has the ability to take over the program from OSM, following the adoption of an approved regulatory program.
7 "Approximate original contour" is defined in section 701(2) of SMCRA as follows:

> The surface configuration achieved by backfilling and grading of the mined area so that the reclaimed area, including any terracing or access roads, closely resembles the general surface configuration of the land prior to mining and blends into and complements the drainage pattern of the surrounding terrain, with all highwalls and spoil piles eliminated; water impoundments may be permitted where the regulatory authority determines that they are in compliance with section 515(b)(8) of the Act.

Federal regulations include a similar definition of AOC at 30 CFR § 701.5.
8 Section 515(e) of SMCRA goes on to provide that such variances are allowed: "provided that the watershed control of the area is improved; and further provided that backfilling with spoil material . . . cover[s] completely the highwall which material will maintain stability following mining and reclamation."
9 See 30 CFR § 740.4.
10 The basic statute for obtaining the property right to conduct metal mining on federal lands is the General Mining Law of 1872 (30 USC §§ 21 et seq.). Obtaining rights to mine on private lands is governed by state property rights law.
11 There is also "low-grade ore," which is set aside or stored for later processing if the operator believes that metal prices might rise at some point in the future, thereby making it profitable to process it.
12 The BLM administers about 264 million acres of land – nearly one-eighth of the United States – mostly in the Western states and in Alaska. The BLM also administers the mineral estate of nearly 700 million acres throughout the country.
13 Rights to mine coal (and various other minerals) on federal lands are obtained by the leasing of the same under the 1920 Mineral Leasing Act and the Federal Coal Leasing Amendments Act of 1976 (30 USC §§ 201 et seq.). Regulatory procedures for leasing federal coal are outlined in 43 CFR Part 3400. Rights to mine gold and other metals on federal lands are generally obtained under the General Mining Law of 1872.
14 See CRS §§ 38-30.5-101 et seq.

Bibliography

30 CFR § 701.5
30 CFR § 740.4
36 CFR Part 228
38 CFR §§ 3809 et seq.
40 CFR Parts 1500 et seq.
43 CFR Part 3400
CRS §§ 34-32-101 et seq.

CRS §§ 38-30.5-101 *et seq.*
16 USC §§ 470 *et seq.*
16 USC §§ 473–78, 479–82, and 551
16 USC §§ 1531 *et seq.*
16 USC § 1532
16 USC §§ 1600–614
30 USC §§ 21 *et seq.*
30 USC §§ 201 *et seq.*
30 USC §§ 1201 *et seq.*
30 USC § 1265b (3)
33 USC §§ 1251 *et seq.*
42 USC §§ 4321 *et seq.*
43 USC §§ 1701 *et seq.*
43 USC §§ 1712 *et seq.*
43 USC § 1732(b).

Index

Abandoned Mine Lands (AML) program 142, 150
acid mine drainage (AMD): Appalachian coal country 64–70, 75; cleanup of 141; University of Virginia at Wise 69–3, 75; Wellington Oro Mine, Colorado 79–84, 121
adaptive approach: landscape design 120–1
alder (*Alnus spp.*) 18, 127
AMD&ART project: Pennsylvania 64–69, 75
Anaconda Smelter, Montana 146
Animas River Stakeholders Group (ARSG) 98–103
Appalachian coal country: acid mine drainage (AMD) 64–70, 75
ArcGIS 116–19, 133
art: and reclamation 64–77
aspen 20, 43

bacteria: reclamation role 18
Bataille, Georges 29, 32, 34
bats: in mine shafts 120
Bear Creek Wind Farm, Pennsylvania 151
Belousov-Zhabotinsky (B-Z) reaction 39
Bénard cells 39
Bingham Canyon Mine, Utah 108
biomass: curves 45–8
black locust (*Robinia pseudoacacia*) 18, 20
Black Mesa Mine, Arizona 133
Breckenridge, Colorado: French Gulch/Wellington Oro mine reclamation 78–86, 121–4; Golden Horseshoe 129–31
brownfield sites: definition 142
Brownfields law 142, 150
Bureau of Land Management (BLM) 158

California Gulch, Colorado 146–7
carbon sequestration 151
"chaorder" 7–9
china-clay extraction: Cornwall 87
Clean Water Act (CWA) 99, 101, 102–3, 142, 160
Clear Creek, Colorado: mining reclamation 57–9
climate change 14
climax ecosystems 48
climbing walls 127
coal mining: acid mine drainage (AMD) 64–77; computer technology 132–9; legislation 155–6
Colorado Water Quality Control Division (WQCD) 99
community involvement: Animas River Stakeholders Group 98–103; Eden Project 96; French Gulch Remediation Opportunities Group 78–86
Comprehensive Environmental Response, Compensation, and Liability Act (CERCLA) 99, 102, 142
computers: mobile 136; software 116–19, 132–9
conservation easements 162
Copper Basin Tailings, Tennessee 144–6
Cornwall: Eden Project 87–97
cottonwood 43
Cranberry Creek Gateway Park, Pennsylvania 147–8
Crowley Creek Collaboration (CCC) 72–7
cryptobiotic soil 127
cyanobacteria 40

databases: United States 56
Deleuze, Gilles 36
digital elevation models (DEM) 53, 59

165

Index

digital simulation: post-mine planning 115–24; trails 131
dogwood (*Cornus spp.*) 20, 127

EarthVision software 137
ecology: of reclamation 16–23, 119–21; science of 44–8; unpredictability 42–3
economic development 147–50
ecosystems: changing 14–16; climax ecosystems 48; development of 45–7; "natural" 42; restoration 144–6; stressed 48–50
ectomycorrhizal fungi (ECM) 17–18
Eden Project, Cornwall 87–97
education: Eden Project 97
Endangered Species Act 160
environmental impact statement (EIS) 159
EPA *see* U.S. Environmental Protection Agency

Federal Brownfields Partnership Mine-Scarred Lands Initiative 148, 150
Federal Land Exchange Facilitation Act 161
Federal Land Policy and Management Act (FLPMA) 158, 161
Federal Mining Dialog 150
Flambeau Mine, Wisconsin 107–8
food chains 44–7
forest fires 48, 131
Forest Service Organic Act 159
French Gulch reclamation: digital simulation 121–4
French Gulch Remediation Opportunities Group (FROG) 78–86
fungi: reclamation role 17–18

gardens: definition 29–30; gold mining reclamation site 30–4
geographic information systems (GIS) 30, 53, 59, 116, 133–4, 137
gold mining: New Zealand 26–34; waste production 29
Golden Horseshoe, Colorado 129–31
Good Samaritan initiatives 102, 150
Guattari, Felix 36

history: preservation of 146–7, 160–1
housing: affordable 84
Hubbard Brook Experimental Forest 49
Hutchinson, G. Evelyn 44

invasive species 15

Kennecott Minerals: sustainable development 105–12

land exchanges 161
Land Revitalization Initiative *see* U.S. Environmental Protection Agency
landscape architecture: adaptive approach 120–1; circulation infrastructure 121; design principles 119–21; digital simulation 115–24; and evolution 40; legislation 154–62; and science 52–60
legislation 103–4, 141–2, 154–62
lodgepole pines 131

McKinley Mine, New Mexico 133, 137, 138
Macraes Flat, New Zealand: Oceana Gold Heritage and Art Park 26–34
maintenance: reclamation sites 20, 131
metals mines: reclamation options 156–7
microorganisms 17–18, 40, 120
Moab, Utah 126
mobile computing 136
modeling software 137
mountain biking trails 126–7
multidisciplinary approaches 52–6, 64–77

National Environmental Policy Act (NEPA) 159–60
National Forest Management Act 159
National Historic Preservation Act 160–1
National Pollutant Discharge Elimination System (NPDES) permits 103
nature: concept of 5; form and function 39
networking: regional 96
New England: ecosystem change 14, 15
New Zealand: Macraes gold mining site 26–34
nitrogen-fixing plants 18

Oceana Gold (NZ): Heritage and Art Park 26–34
Ocoee River, Tennessee 144–6
Odum, Eugene P. 44–7
Office of Surface Mining (OSM): Technical Innovation and Professional Services (TIPS) program 132–9
Olmsted, Frederick Law 23
open space: preservation 84–6
open-pit mining: pre-mine planning 125–8
order and disorder 6–9, 37
organic matter "recycle" 17

P/B ratio 47–8
partnerships: post-mining regeneration 96
paths *see* trails
Pennsylvania: AMD&ART project 64–70, 75
Phragmites australis (common reed) 15–16
Pisolithus tinctorius ("Pt") 17

Index

plants: choice for reclamation sites 18–23, 127; disturbance-adapted 120
post-mine planning: French Gulch Simulation Project 115, 121–4; Kennecott Minerals 105–12
Post-Mining Alliance: Eden Project 87–8, 93–7
Project for Reclamation Excellence (P-REX): French Gulch/Wellington Oro reclamation 115, 121–4; reuse of haul roads 125–8

reclamation: definition 13; digital simulation 115–24; ecology of 16–23, 119–21
recreational uses: golf course 148; trails 66–8, 123–4, 126–7, 129–31, 146–7
reeds 15–16
regulations *see* legislation
remediation: definition 5
remote sensing 137–9
renewable energy facilities 148–50, 151
restoration: definition 13
Rio Tinto: Eden Project 91, 93; Kennecott Minerals 106, 108
riparian corridors 126–8
roads: reuse of haul roads 125–8; *see also* trails

Salt Lake Valley: Daybreak Project 109–12
science: and landscape architecture 52–6
Silver Bow Creek, Montana 144, 152
Sitka Center for Art and Ecology 72–5
software: landscape design 116–19; mine-permit applications 132–9
succession: process of 13–14, 43–4
Summitville Mine, Colorado 77, 157, 158
Sunnyside Gold Corporation (SGC) 101–2
Superfund Program *see* U.S. Environmental Protection Agency
Surface Mining Control and Reclamation Act (SMCRA) 144, 155–6
sustainable development: definition 105–6; Kennecott Minerals 105–12

thermodynamics 7–9, 29, 37–9, 42
Thoreau, Henry David 9–11, 43–4
tourism: Eden Project 87–97; French Gulch 78, 84–5; Oceana Gold Heritage and Art Park 26–34; post-mining areas 95–6

trails: California Gulch 146–7; French Gulch/Wellington Oro site 121, 123–4; Ghost Town Rail Trail 66–8; Golden Horseshoe 129–31; mountain biking 126–7; zoning 130; *see also* roads
trout fisheries: French Creek 81–4

University of Virginia at Wise (UVA-Wise): acid mine drainage (AMD) 69–3, 75
urban areas: ecological restoration 14–16
U.S. Abandoned Mine Lands (AML) program 99, 142, 150
U.S. Environmental Protection Agency (EPA): Land Revitalization Initiative 141–52; performance measures 150–1; Superfund Program 77, 78, 99, 142
U.S. Forest Service: legislation 159
U.S. Geological Survey (USGS): information databases 56; landscape architecture 56–9
Use Attainability Analysis: Animas River watershed 99, 101
Utah: Daybreak Project 108–12
Utah Copper Company 108

vegetation *see* plants
vesicular-arbuscular mycorrhizae (VAM) 17–18
Vintondale, Pennsylvania: AMD&ART project 65–70, 75

waste: adaptation to life 40; definition 29
water quality: Animas River 99–103; *see also* acid mine drainage
watershed restoration: Animas River 98–103; Crowley Creek Collaboration (CCC) 72–7
Wellington Oro Mine, Colorado: reclamation 78–86, 121–4
wetlands: restoration 151–2
wildlife: protection of 160
willow (*Salix spp.*) 122, 127
Wisconsin: Flambeau Mine 106–7
Wyoming: coal mining 134, 136, 137

Yellowstone National Park 43, 48

zinc: as indicator element 102

eBooks – at www.eBookstore.tandf.co.uk

A library at your fingertips!

eBooks are electronic versions of printed books. You can store them on your PC/laptop or browse them online.

They have advantages for anyone needing rapid access to a wide variety of published, copyright information.

eBooks can help your research by enabling you to bookmark chapters, annotate text and use instant searches to find specific words or phrases. Several eBook files would fit on even a small laptop or PDA.

NEW: Save money by eSubscribing: cheap, online access to any eBook for as long as you need it.

Annual subscription packages

We now offer special low-cost bulk subscriptions to packages of eBooks in certain subject areas. These are available to libraries or to individuals.

For more information please contact webmaster.ebooks@tandf.co.uk

We're continually developing the eBook concept, so keep up to date by visiting the website.

www.eBookstore.tandf.co.uk